Holt McDougal Mathematics

Course 2
Chapter 4 Resource Book

HOLT McDOUGAL
a division of Houghton Mifflin Harcourt

Printed in the United States of America

ISBN 13: 978-0-55-400731-1
ISBN 10: 0-55-400731-2

2 3 4 5 170 12 11 10 09

Contents

Section 4A Family Letter 1
Section 4A At-Home Practice 3
Section 4A Family Fun 4
4-1 Practice A, B, C 5
4-1 Review for Mastery 8
4-1 Challenge 10
4-1 Problem Solving 11
4-1 Reading Strategies 12
4-1 Puzzles, Twisters & Teasers 13
4-2 Practice A, B, C 14
4-2 Review for Mastery 17
4-2 Challenge 18
4-2 Problem Solving 19
4-2 Reading Strategies 20
4-2 Puzzles, Twisters & Teasers 21
4-3 Practice A, B, C 22
4-3 Review for Mastery 25
4-3 Challenge 26
4-3 Problem Solving 27
4-3 Reading Strategies 28
4-3 Puzzles, Twisters & Teasers 29
4-4 Practice A, B, C 30
4-4 Review for Mastery 33
4-4 Challenge 35
4-4 Problem Solving 36
4-4 Reading Strategies 37
4-4 Puzzles, Twisters & Teasers 38
Section 4B Family Letter 39
Section 4B At-Home Practice 41
Section 4B Family Fun 42
4-5 Practice A, B, C 43
4-5 Review for Mastery 46
4-5 Challenge 47
4-5 Problem Solving 48
4-5 Reading Strategies 49
4-5 Puzzles, Twisters & Teasers 50
4-6 Practice A, B, C 51
4-6 Review for Mastery 54
4-6 Challenge 55
4-6 Problem Solving 56
4-6 Reading Strategies 57
4-6 Puzzles, Twisters & Teasers 58
4-7 Practice A, B, C 59
4-7 Review for Mastery 62
4-7 Challenge 63
4-7 Problem Solving 64
4-7 Reading Strategies 65
4-7 Puzzles, Twisters & Teasers 66
Section 4C Family Letter 67
Section 4C At-Home Practice 69
Section 4C Family Fun 70
4-8 Practice A, B, C 71
4-8 Review for Mastery 74
4-8 Challenge 75
4-8 Problem Solving 76
4-8 Reading Strategies 77
4-8 Puzzles, Twisters & Teasers 78
4-9 Practice A, B, C 79
4-9 Review for Mastery 82
4-9 Challenge 83
4-9 Problem Solving 84
4-9 Reading Strategies 85
4-9 Puzzles, Twisters & Teasers 86
4-10 Practice A, B, C 87
4-10 Review for Mastery 90
4-10 Challenge 91
4-10 Problem Solving 92
4-10 Reading Strategies 93
4-10 Puzzles, Twisters & Teasers 94
Answers 95

Description of Contents

Family Involvement Pages

The Chapter Resource Book includes a set of family involvement pages for each section. The family involvement pages consist of the following items:

- **Family Letter**, a two-page letter describing the math that the student will study in each section. A list of vocabulary words from the lessons is included, and some worked-out examples are provided.
- **At-Home Practice**, a one-page worksheet with problems drawn from the content described in the Family Letter. Answers are provided on the page so that the student and his or her family can check this work at home.
- **Family Fun**, a one-page activity sheet that the student and his or her family can work on together.

Practice A, B, and C

There are three practice worksheets for every lesson. All of these reinforce the content of the lesson. Practice B is shown in the Teacher's Edition and is appropriate for the on-level student. It is also available as a workbook (Homework & Practice Workbook)

Practice A is easier than Practice B but still practices the content of the lesson. Practice C is more challenging than Practice B.

Review for Mastery

The Review for Mastery worksheet (one per lesson) provides an alternate way to teach or review the main concepts of the lesson. This worksheet is one or two pages long and is shown in the Teacher's Edition.

Challenge

The Challenge worksheet (one per lesson) enhances critical thinking skills and extends the lesson. This worksheet is shown in the Teacher's Edition.

Problem Solving

The Problem Solving worksheet (one per lesson) provides practice in problem solving and opportunities for real-world applications and for interdisciplinary connections. There are both multiple choice and short response problems. This worksheet is shown in the Teacher's Edition.

Reading Strategies

The Reading Strategies worksheet (one per lesson) provides tools to help the student master math vocabulary or symbols.

Puzzles, Twisters & Teasers

The Puzzles, Twisters & Teasers worksheet (one per lesson) provides fun practice while reinforcing the content of the lesson.

CHAPTER 4

Family Letter

4A Ratios, Rates, and Proportions

Dear Family,

The student will be learning about ratios, rates, and proportions. A **ratio** compares two quantities using division. A ratio can be written in three different ways.

The local library has 15 part-time workers and three full-time librarians. Write the ratio in all three forms.

Part-time workers to full-time librarians: $\frac{15}{3}$ 15:3 15 to 3

A ratio that compares two values with different units is called a **rate**. The **unit rate** compares a quantity to a unit of one.

Chas read 135 pages at a constant rate in 5 hours. How many pages did he read per hour? Write the unit rate in fraction and word form.

$\frac{135 \text{ pages}}{5 \text{ hours}}$ Compare pages and hours.

$\frac{135 \div 5}{5 \div 5} = \frac{27}{1}$ Divide the numerator and denominator by 5.

Chas read 27 pages per hour.

An equation stating that two ratios are equivalent is called a **proportion**. You can determine whether two ratios are equivalent by writing the ratios in simplest form. If the ratios can be simplified to the same fraction, then the ratios are in proportion.

Determine whether the ratios $\frac{5}{12}$ and $\frac{15}{36}$ are proportional.

$\frac{5}{12}$ $\frac{5}{12}$ is already in simplest form.

$\frac{15 \div 3}{36 \div 3} = \frac{5}{12}$ Simplify $\frac{15}{36}$.

Since $\frac{5}{12} = \frac{5}{12}$, the ratios are proportional.

Vocabulary

These are the math words we are learning:

cross product for two ratios, the product of the first term in one ratio and the second term in the other ratio

equivalent ratios ratios that name the same comparison

proportion an equation that says two ratios are equivalent

rate a ratio that compares two quantities measured in different units

ratio a comparison of two quantities by division

unit rate a rate whose denominator is 1 when written as a fraction

CHAPTER 4

Family Letter

4A Ratios, Rates, and Proportions *continued*

The student will use **cross products** to solve proportions.

> *In the proportion* $\frac{a}{b} = \frac{c}{d}$, *where* $b \neq 0$ and $d \neq 0$,
> *the cross products,* $a \cdot d$ and $b \cdot c$, *are equal.*

Use cross products to solve the proportion.

$$\frac{3}{15} = \frac{5}{t}$$

$3 \cdot t = 15 \cdot 5$ The cross products are equal.

$3t = 75$ Multiply.

$\frac{3t}{3} = \frac{75}{3}$ Divide each side by 3.

$t = 25$

It is important to set up proportions correctly. Each ratio must compare corresponding quantities in the same order. Suppose a boat travels 16 miles in 4 hours and 8 miles in x hours at the same speed. Either of these proportions could represent this situation.

Trip 1 → $\frac{16 \text{ mi}}{4 \text{ hr}} = \frac{8 \text{ mi}}{x \text{ hr}}$ ← Trip 2

$\frac{16 \text{ mi}}{8 \text{ mi}} = \frac{4 \text{ hr}}{x \text{ hr}}$ (top row ← Trip 1; bottom row ← Trip 2)

The student will be able to apply these skills to real-life scenarios. Encourage the student to practice solving proportions.

Sincerely,

Name ______________________ Date ______________ Class ______________

CHAPTER 4

At-Home Practice

4A Ratios, Rates, and Proportions

A school has 42 sixth-graders, 56 seventh-graders, and 49 eighth-graders. Write each ratio in all three forms.

1. sixth-graders to eighth-graders ______________________

2. seventh-graders to total students ______________________

Find the unit rates and write them in both fraction and word form.

3. Rachel drives 180 miles in 3 hours. ______________________

4. A store sells 12 drinks for $3.60. ______________________

5. 150 people travel in 3 buses ______________________

6. 250 cookies in 10 packages ______________________

Determine whether the ratios are proportional.

7. $\frac{12}{18}, \frac{20}{30}$ ______________

8. $\frac{25}{40}, \frac{15}{24}$ ______________

9. $\frac{28}{49}, \frac{16}{36}$ ______________

Use cross products to solve each proportion.

10. $\frac{6}{x} = \frac{21}{28}$ ______________

11. $\frac{5}{7} = \frac{m}{28}$ ______________

12. $\frac{8}{24} = \frac{6}{t}$ ______________

13. $\frac{3}{15} = \frac{h}{35}$ ______________

14. $\frac{18}{30} = \frac{f}{10}$ ______________

15. $\frac{j}{27} = \frac{28}{36}$ ______________

Answers: 1. $\frac{6}{7}$, 6 to 7, 6:7 **2.** $\frac{8}{21}$, 8 to 21, 8:21 **3.** $\frac{60 \text{ mi}}{1 \text{ hr}}$; Rachel drives 60 miles in 1 hour. **4.** $\frac{\$0.30}{1 \text{ drink}}$; A store sells 1 drink for $0.30. **5.** $\frac{50 \text{ people}}{1 \text{ bus}}$; 50 people travel on one bus. **6.** $\frac{25 \text{ cookies}}{1 \text{ package}}$; A package has 25 cookies. **7.** yes; $\frac{2}{3} = \frac{2}{3}$ **8.** yes; $\frac{5}{8} = \frac{5}{8}$ **9.** no; $\frac{4}{7} \neq \frac{4}{9}$ **10.** $x = 8$ **11.** $m = 20$ **12.** $t = 18$ **13.** $h = 7$ **14.** $f = 6$ **15.** $j = 21$

Name ______________________ Date ______________ Class ______________

CHAPTER 4

Family Fun

Go Fish!

Directions

- Cut out the cards, shuffle, and give each player 3 cards. First determine if you have a match. A proportion and its solution are considered a match.
- If you do not have a matched pair, ask for a card from your opponent's hand. If your opponent has the card you need, he or she must give you that card. If your opponent does not have the card that you requested, then you fish from the rest of the deck.
- You can use the bonus card only for a proportion, not the solution. To use the bonus card, you must create a proportion and ask for the solution card.
- The winner is the first player to get 3 matches.

$\frac{2}{6} = \frac{x}{21}$	7	$\frac{x}{8} = \frac{3}{6}$	4	$\frac{4}{7} = \frac{12}{x}$
21	$\frac{15}{5} = \frac{x}{4}$	12	$\frac{16}{24} = \frac{2}{x}$	3
$\frac{5}{15} = \frac{x}{18}$	6	$\frac{5}{8} = \frac{x}{24}$	15	$\frac{5}{x} = \frac{3}{27}$
45	$\frac{13}{26} = \frac{x}{4}$	2	$\frac{7}{14} = \frac{4}{x}$	8
$\frac{10}{120} = \frac{1}{x}$	12	$\frac{6}{21} = \frac{4}{x}$	14	**Bonus**

Name ______________________ Date ______________ Class ______________

LESSON 4-1

Practice A

Ratios

Match the ratios.

A farmer has 5 pigs, 13 chickens, and 8 cows.

1. cows to pigs — 13:5
2. chickens to pigs — 5:8
3. cows to chickens — 8:5
4. pigs to cows — 8:13

The school orchestra has 9 cellos, 14 flutes, and 17 violins.

5. cellos to violins — 9 to 14
6. flutes to cellos — 9 to 17
7. violins to flutes — 17 to 14
8. cellos to flutes — 14 to 9

Miguel has 8 pennies, 5 nickels, and 3 quarters.

9. nickels to pennies — $\frac{3}{8}$
10. pennies to quarters — $\frac{5}{3}$
11. nickels to quarters — $\frac{5}{8}$
12. quarters to pennies — $\frac{8}{3}$

A bowl has 16 grapes, 7 cherries, and 9 strawberries.

13. grapes to strawberries — 7:16
14. cherries to grapes — 7:9
15. strawberries to cherries — 16:9
16. cherries to strawberries — 9:7

A baseball team has 4 pitchers, 10 outfielders, and 12 infielders. Write each ratio in all three forms.

17. pitchers to infielders

____

18. infielders to outfielders

____

19. pitchers to outfielders

____

20. outfielders to entire team

____

21. Meg used 24 red tiles and 12 yellow tiles to make a design. Write the ratio of red tiles to yellow tiles in simplest form.

____

22. Tell which club has the greater ratio of girls to boys.

____

	Movie Club	Art Club
Girls	16	8
Boys	14	4

Name ______________________ Date ______________ Class ______________

LESSON 4-1

Practice B

Ratios

The annual dog show has 22 collies, 28 boxers, and 18 poodles. Write each ratio in all three forms.

1. collies to poodles

____

2. boxers to collies

____

3. poodles to boxers

____

4. poodles to collies

____

The Franklin School District has 15 art teachers, 27 math teachers, and 18 Spanish teachers. Write the given ratio in all three forms.

5. art teachers to math teachers

____

6. math teachers to Spanish teachers

____

7. Spanish teachers to all teachers

____

8. art and math teachers to Spanish teachers

____

9. Thirty-two students are asked whether the school day should be longer. Twenty-four vote "no" and 8 vote "yes." Write the ratio of "no" votes to "yes" votes in simplest form.

____

10. A train car has 64 seats. There are 48 passengers on the train. Write the ratio of seats to passengers in simplest form.

____

11. Tell whose CD collection has the greater ratio of rock CDs to total CDs.

____

	Glen	Nina
Classical CDs	4	8
Rock CDs	9	12
Other CDs	5	7

Name ____________________ Date ____________ Class ____________

LESSON 4-1

Practice C

Ratios

A traveling theater company has 51 actors, 63 stagehands, and 27 set designers. Write each ratio in all three forms.

1. stagehands to set designers
____

2. actors to stagehands
____

3. set designers to actors
____

4. set designers to stagehands and actors
____

There are 18 monkeys, 6 gorillas, and 15 other apes in the Primate House at the zoo. Write different combinations of animals that have the following ratios.

5. 2:5 _______________

6. 6:5 _______________

7. 1:3 _______________

8. 6:7 _______________

9. 5:13 _______________

10. 8:5 _______________

11. The class library has 7 French books, 11 Spanish books, 19 art books, 3 Italian books, and 14 science books. What is the ratio of foreign language books to all books in simplest form? __________

12. A market research company tests new commercials. Out of 96 people, 80 people prefer Commercial A and the rest prefer Commercial B. Write the ratio in simplest form of people who prefer Commercial A to Commercial B.
____

13. Which grade has the greatest ratio of the number of students against wearing a uniform to the total number of students in that grade?
____

School Uniform Survey

	6th Grade	7th Grade	8th Grade
For	16	9	5
Against	20	24	21
No Opinion	4	3	1

Name _______________________ Date _______________ Class _______________

LESSON 4-1

Review for Mastery

Ratios

A **ratio** is a comparison of two numbers.

Tamara has 2 dogs and 8 fish. The ratio of dogs to fish can be written in three different ways.

	Ratio	Ratio in simplest form
• using the word *to*	2 to 8	1 to 4
• using a colon (:)	2:8	1:4
• writing a fraction	$\frac{2}{8}$	$\frac{1}{4}$

You can read the ratios as *2 to 8* or *1 to 4*.

In a basket of fruit, there are 8 apples, 3 bananas, and 5 oranges. Write each ratio in all three forms.

1. apples to bananas

   There are _____ apples and _____ bananas. So, the ratio of apples to bananas is _____ to _____, or _____ : _____, or —.

2. oranges to apples

   There are _____ oranges and _____ apples. So, the ratio of oranges to apples is _____ to _____, or _____ : _____, or —.

3. bananas to oranges

   There are _____ bananas and _____ oranges. So, the ratio of bananas to oranges is _____ to _____, or _____ : _____, or —.

4. apples to all pieces of fruit

   There are _____ apples and _____ pieces of fruit in all. So, the ratio of apples to all pieces of fruit is _____ to _____, or _____ : _____, or —.

A large bouquet of flowers is made up of 18 roses, 16 daisies, and 24 iris. Write each ratio in all three forms.

5. roses to iris

   ____

6. daisies to roses

   ____

7. iris to daisies

   ____

8. roses to all flowers

   ____

Name ______________________ Date ______________ Class ______________

LESSON 4-1

Review for Mastery

Ratios (continued)

To compare ratios, write them as fractions with a common denominator. Then compare the numerators.

Tell whose bag has the greater ratio of solid marbles to striped marbles.

	Ken	Val
Solid	5	7
Striped	9	12

Step 1: Write the ratio of solid marbles to striped marbles for each bag. Write each ratio as a fraction.

Ken's Bag: $\frac{\text{Solid}}{\text{Striped}} = \frac{5}{9}$

Val's Bag: $\frac{\text{Solid}}{\text{Striped}} = \frac{7}{12}$

Step 2: Choose a common denominator.

36 is a common denominator for 9 and 12.

Step 3: Write each fraction using the common denominator.

$\frac{5}{9} = \frac{5 \times 4}{9 \times 4} = \frac{20}{36}$ $\quad \frac{7}{12} = \frac{7 \times 3}{12 \times 3} = \frac{21}{36}$

Step 4: Compare the numerators.

$20 < 21$

Since $20 < 21$, $\frac{20}{36} < \frac{21}{36}$ and $\frac{5}{9} < \frac{7}{12}$. So, Val's bag has the greater ratio of solid marbles to striped marbles.

9. Tell whose bookshelf has the greater ratio of novels to biographies.

	Tina	Mark
Novels	3	5
Biographies	5	8

For Tina's bookshelf: $\frac{\text{Novels}}{\text{Biographies}} = \frac{\quad}{\quad}$

For Marks's bookshelf: $\frac{\text{Novels}}{\text{Biographies}} = \frac{\quad}{\quad}$

Write the fractions so they have a common denominator. Then compare the numerators.

____

__________'s bookshelf has the greater ratio of novels to biographies.

Name ____________________ Date ____________ Class ____________

LESSON 4-1

Challenge

The Golden Ratio

The **Golden Ratio**, which is also known as the Golden Section, was given its name because architects and artists throughout history have used it to produce shapes that are pleasing to look at. They have used this ratio as the ratio of length to width in various shapes.

The Golden Ratio can be represented by a line segment that meets these conditions:

- The line segment is divided into two sections of different lengths.
- The ratio of the longer section to the shorter section is equal to the ratio of the whole segment to the longer section

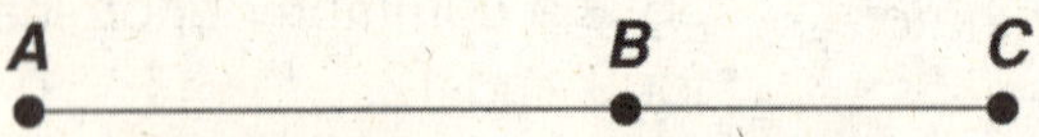

$$\frac{\text{longer?}(AB)}{\text{shorter?}(BC)} = \frac{\text{whole?}(AC)}{\text{longer?}(AB)}$$

Find each measurement to the nearest millimeter. Use your measurements to write the given ratio. Then write the ratio as a decimal rounded to the nearest hundredth.

1. 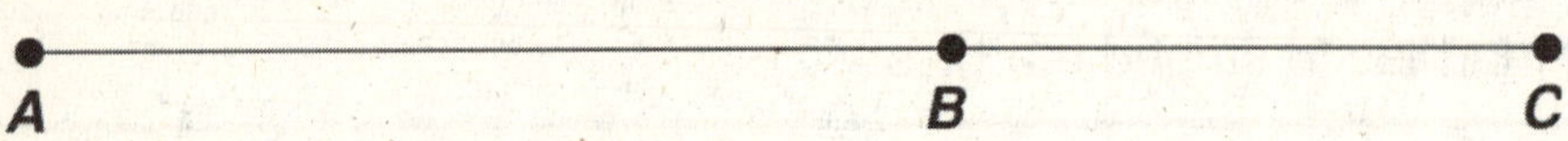

$\frac{AB}{BC} = \frac{\_\_}{\_\_} =$ ________ $\frac{AC}{AB} = \frac{\_\_}{\_\_} =$ ________

2. 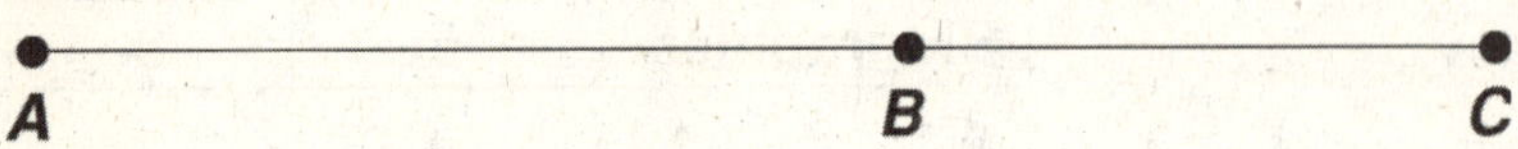

$\frac{AB}{BC} = \frac{\_\_}{\_\_} =$ ________ $\frac{AC}{AB} = \frac{\_\_}{\_\_} =$ ________

3. 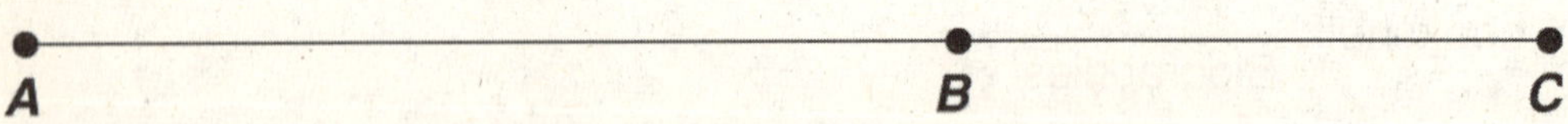

$\frac{AB}{BC} = \frac{\_\_}{\_\_} =$ ________ $\frac{AC}{AB} = \frac{\_\_}{\_\_} =$ ________

4. In which exercise is the line segment divided so that it is closest to representing the golden ratio? Explain.

____

____

Name ______________________ Date ____________ Class ____________

LESSON 4-1

Problem Solving

Ratios

Write the correct answer.

1. The Rockport Diner has 8 seats at the counter and 32 seats at tables. Of these seats, 16 are taken. Write the ratio of seats taken to empty seats in simplest form three ways.

____

2. During the 2001 WNBA season, the Los Angeles Sparks had 28 wins and only 4 losses. Write the ratio of wins to games played in simplest form three ways.

____

3. For every 300 people surveyed in 2002, 186 said their favorite Winter Olympic sport was figure skating. Write this ratio in simplest form three ways.

____

4. In 2004, George W. Bush received 286 electoral votes, and John Kerry received 251, and 1 elector voted for John Edwards. Write the ratio of Bush's electoral votes to total electoral votes in simplest form three ways.

____

Choose the letter for the best answer.

5. There are 62 girls in the seventh grade and 58 boys in the eighth grade. Each grade has 120 students. Which statement correctly compares the ratios of boys to girls in each grade?

   A The eighth-grade ratio is greater.

   B The seventh-grade ratio is greater.

   C The eighth-grade ratio is lesser.

   D Both ratios are equal.

6. Matt has 6 video racing games and 8 video sports games. Which ratio is the ratio of racing games to total video games in simplest form?

   F $\frac{3}{4}$

   G $\frac{3}{7}$

   H $\frac{4}{3}$

   J $\frac{4}{7}$

7. Which player has the greatest ratio of baskets to total shots?

   A Marisol

   B Nina

   C Joanne

   D Talia

	Baskets	**Missed Shots**
Marisol	8	8
Nina	7	5
Joanne	2	4
Talia	5	3

Name ______________________ Date ______________ Class ______________

LESSON 4-1

Reading Strategies

Build Vocabulary

A **ratio** is a way to compare two numbers.

The ratio of black counters to white counters is three to two.

There are three ways to write ratios:

3 to 2 ← read "3 to 2"

3:2 ← read "3 to 2"

$\frac{3}{2}$ ← read "3 to 2"

Now compare the white counters to the black counters.

1. Use a fraction to write this ratio.

____

2. Write the ratio two other ways.

____

A team has 5 boys and 4 girls. Use this information to complete Exercises 3–6.

3. Write a ratio as a fraction to compare the number of boys to the number of girls.

____

4. Write this ratio two other ways.

____

5. Use division to compare the number of girls to the total number of players on the team.

____

6. Write this ratio two other ways.

____

Name ______________________ Date ______________ Class ______________

LESSON 4-1

Puzzles, Twisters & Teasers

Find Your Ratio

Use the letters in the word RATIO to write a fraction for each ratio in simplest form.

RATIO

1. number of Ts:number of Rs ________

2. consonants:vowels ________

3. vowels:total number of letters ________

4. consonants:total number of letters ________

Next, use the letters in the word MATHEMATICS to write the fraction for each ratio in simplest form.

5. consonants:total number of letters ________

6. vowels:consonants ________

7. number of Ms:total number of letters ________

8. number of Hs:total number of letters ________

9. number of Ms and Ts:total number of letters ________

Fraction Box

$A = \frac{1}{7}$

$D = \frac{3}{5}$

$E = \frac{2}{5}$

$G = 1$

$H = \frac{1}{11}$

$I = \frac{2}{11}$

$L = \frac{2}{7}$

$M = \frac{5}{7}$

$O = \frac{2}{3}$

$R = \frac{6}{7}$

$S = \frac{4}{7}$

$T = \frac{4}{11}$

$Y = \frac{7}{11}$

Find the letters that match your answers in the Fraction Box. To solve the riddle, write the letters on the blanks that correspond to the problem numbers.

What do you need to spot an iceberg 20 miles away?

___ ___ ___ ___
1 2 2 3

___ ___ ___ ___ ___ ___ ___ ___
4 5 4 6 7 1 8 9

Name ______________________ Date ______________ Class ______________

LESSON 4-2

Practice A

Rates

1. To make 2 batches of brownies, Ed needs. 4 eggs. How many eggs are needed per batch of brownies? ______________

$$\frac{4 \text{ eggs}}{2 \text{ batches}} = \frac{\_\_\_ \text{ eggs}}{1 \text{ batch}}$$

2. Jenny drives 265 miles in 5 hours. What is her average rate of speed in miles per hour? ______________

$$\frac{265 \text{ miles}}{5 \text{ hours}} = \frac{\_\_\_ \text{ miles}}{1 \text{ hour}}$$

3. A job pays $56 for 8 hours of work. How much money does the job pay per hour? ______________

4. Ned scores 84 points in 6 games. How many points per game does Ned score? ______________

5. A 6-ounce blueberry muffin has 450 calories. How many calories are there per ounce? ______________

6. A parking garage charges $21 for 6 hours. How much does the garage charge per hour? ______________

7. The Rylands want to drive 360 miles in 8 hours. What is their average speed in miles per hour? ______________

8. A plane travels 2,395 miles in 5 hours. What is the plane's average speed in miles per hour? ______________

9. A 16-ounce bottle of fruit punch costs $2.40. A 24-ounce bottle of fruit punch costs $3.84. Which size costs less per ounce? ______________

$$\frac{\$2.40}{16 \text{ oz}} = \frac{\$\_\_\_}{1 \text{ oz}} \qquad \frac{\$3.84}{24 \text{ oz}} = \frac{\$\_\_\_}{1 \text{ oz}}$$

Name ______________________ Date ______________ Class ______________

LESSON 4-2

Practice B

Rates

1. A part-time job pays $237.50 for 25 hours of work. How much money does the job pay per hour? ______________

2. A class trip consists of 84 students and 6 teachers. How many students per teacher are there? ______________

3. A factory builds 960 cars in 5 days. What is the average number of cars the factory produces per day? ______________

4. The Wireless Cafe charges $5.40 for 45 minutes of Internet access. How much money does The Wireless Cafe charge per minute? ______________

5. A bowler scores 3,152 points in 16 games. What is his average score in points per game? ______________

6. Melissa drives 238 miles in 5 hours. What is her average rate of speed? ______________

7. An ocean liner travels 1,233 miles in 36 hours. What is the ocean liner's average rate of speed? ______________

8. A plane is scheduled to complete a 1,792-mile flight in 3.5 hours. In order to complete the trip on time, what should be the plane's average rate of speed? ______________

9. The Nuthouse sells macadamia nuts in three sizes. The 12 oz jar sells for $8.65, the 16 oz jar sells for $10.99, and the 24 oz gift tin costs $16.99. Which size costs less per ounce? ______________

Name ______________________ Date ______________ Class ______________

LESSON 4-2

Practice C

Rates

1. Maria earns $603.75 for 35 hours of work. What is her rate of pay per hour? ______________

2. The Ranch House serves a 24 oz sirloin steak that has a total of 1,800 calories. How many calories per ounce does the steak have? ______________

3. Kia stuffs 228 envelopes in an hour. What is the average number of envelopes Kia stuffs per minute? ______________

4. A freight train travels 1445 miles in 25 hours. What is the train's average rate of speed? ______________

5. June runs 600 yards in 2 min 5 sec. What is her average speed? ______________

6. A plane travels 294 miles in 45 minutes. What is its average speed in miles per hour? ______________

7. A pitcher's earned run average is the number of earned runs allowed per game of 9 innings. A pitcher allows 20 earned runs in 50 innings. What is the pitcher's earned run average? ______________

Find each unit price. Then decide which has the lowest price per unit.

8. $\frac{\$6.48}{36\text{ oz}}$ or $\frac{\$8.16}{48\text{ oz}}$

____

____

9. $\frac{\$9.03}{7\text{ ft}}$ or $\frac{\$15.84}{12\text{ ft}}$

____

____

10. $\frac{\$25.35}{6.5\text{ lb}}$ or $\frac{\$31.60}{8\text{ lb}}$

____

____

11. $\frac{\$9.16}{0.4\text{ m}}$ or $\frac{\$13.20}{0.6\text{ m}}$

____

____

Name ______________________ Date ____________ Class ____________

LESSON 4-2

Review for Mastery

Rates

A **rate** is a ratio that compares two different kinds of quantities or measurements. Rates can be simplified. Rates sometimes use the words *per* and *for* instead of *to*, such as 55 miles per hour and 3 tickets for $1.

The scale on a map might be 3 inches equals 60 miles. Simplify by dividing both numerator and denominator by the same number.

$$\frac{3 \text{ inches}}{60 \text{ miles}} = \frac{3 \div 3}{60 \div 3} = \frac{1 \text{ inch}}{20 \text{ miles}}, \text{ or 1 inch to 20 miles}$$

A **unit rate** is a rate per 1 unit. So, in a unit rate, the denominator is always 1.

Miguel can type 180 words in 4 minutes.

$$\frac{180 \text{ words}}{4 \text{ minutes}} = \frac{180 \div 4}{4 \div 4} = \frac{45 \text{ words}}{1 \text{ minute}}, \text{ or 45 words per minute}$$

↑ rate ↑ unit rate ↑ word form

Find each unit rate. Write in both fraction and word form.

1. Film costs $7.50 for 3 rolls.

$$\frac{\$7.50}{3 \text{ rolls}} = \frac{7.50 \div}{3 \div} = \frac{\$\_\_\_\_}{1 \text{ roll}}$$

Word form: ______________________

2. Drive 288 miles on 16 gallons of gas.

$$\frac{288 \text{ mi}}{16 \text{ gal}} = \frac{288 \div}{16 \div} = \frac{\_\_\_\_ \text{ mi}}{1 \text{ gal}}$$

Word form: ______________________

3. Earn $52 for 8 hours of work.

$$\frac{\$52.00}{8 \text{ hr}} = \frac{52 \div}{8 \div} = \frac{\$\_\_\_\_}{\_\_\_\_ \text{ hr}}$$

Word form: ______________________

4. Use 5 quarts of water for every 2 pounds of chicken.

$$\frac{5 \text{ qt}}{2 \text{ lb}} = \frac{5 \div}{2 \div} = \frac{\_\_\_\_ \text{ qt}}{\_\_\_\_ \text{ lb}}$$

Word form: ______________________

5. Snowfall of 12 inches in 4 hours

____

____

6. 90 students and 5 teachers

____

____

Name ______________________ Date ______________ Class ______________

LESSON 4-2

Challenge

The Rate Maze

Contestants A–H must make their way through the maze to the Winners' Circle. To reach the Winners' Circle, each contestant must find a path from his or her current location through sections containing equivalent unit prices or rates. The contestants can move only through sections that share a corner.

Find each unit price or rate. Circle the two contestants who will *not* be able to get to the Winners' Circle.

Name ______________________ Date ______________ Class ______________

LESSON 4-2

Problem Solving

Rates

Write the correct answer.

1. A truck driver drives from Cincinnati to Boston in 14 hours. The distance traveled is 840 miles. What is the truck driver's average rate of speed?

____

2. Melanie earns $97.50 in 6 hours. Earl earns $296.00 in 20 hours. Who earns a higher rate of pay per hour?

____

3. Mr. Tanney buys a 10-trip train ticket for $82.50. Ms. Elmer buys an unlimited weekly pass for $100 and uses it for 12 trips during the week. Write the unit cost per trip for each person.

____

____

4. Metropolitan Middle School has 564 students and 24 teachers. Eastern Middle School has 623 students and 28 teachers. Which school has the lower unit rate of students per teacher?

____

Choose the letter for the best answer.

5. Which shows 20 pounds per 5 gallons as a unit rate?

   A $\frac{20 \text{ lb}}{1 \text{ gal}}$ C $\frac{5 \text{ lb}}{1 \text{ gal}}$

   B $\frac{4 \text{ lb}}{1 \text{ gal}}$ D $\frac{1 \text{ gal}}{4 \text{ lb}}$

6. What is the unit price of a 6-ounce tube of toothpaste that costs $3.75?

   F $0.06

   G $0.23

   H $0.62

   J $0.63

7. Max bought 16 gallons of gas for $40.64. Lydia bought 12 gallons of gas for $31.08. Kesia bought 18 gallons of gas for $45.72. Who got the best buy?

   A Max got the best buy.

   B Lydia got the best buy.

   C Lydia and Kesia both paid the same rate, which is better than Max's rate.

   D Max and Kesia both paid the same rate, which is better than Lydia's rate.

8. A pack of 12 8-ounce bottles of water costs $3.36. What is the unit cost per ounce of bottled water?

   F $0.03 per ounce

   G $0.04 per ounce

   H $0.28 per bottle

   J $0.42 per bottle

Name ______________________ Date ______________ Class ______________

LESSON 4-2

Reading Strategies

Build Vocabulary

A **rate** is a special ratio that compares two values that are measured in different units.

- $8 for 2 pounds of beef → $\frac{\$8}{2 \text{ lb}}$ → pounds compared to dollars
- 12 miles in 3 hours → $\frac{12 \text{ mi}}{3 \text{ h}}$ → miles compared to hours

1. Is the ratio $\frac{5 \text{ hours}}{12 \text{ hours}}$ a rate? Explain.

____

2. Is the ratio $\frac{50 \text{ yards}}{18 \text{ seconds}}$ a rate? Explain.

____

3. What does the rate $\frac{375 \text{ miles}}{15 \text{ gallons}}$ compare?

____

In a **unit rate**, the second quantity in the rate is 1 unit. To write a unit rate, write the rate as a fraction with a denominator of 1.

$$\frac{\$8}{2 \text{ lb}} = \frac{(\$8 \div 2)}{(2 \text{ lb} \div 2)} = \frac{\$4}{1 \text{ lb}}$$

↑ rate ↑ unit rate

Write *yes* or *no* to tell if each rate is a unit rate. If it is not a unit rate, write the unit rate.

4. $\frac{\$2.75}{1 \text{ h}}$ ____________

5. $\frac{100 \text{ mi}}{4 \text{ gal}}$ ____________

6. $\frac{35 \text{ lb}}{1 \text{ box}}$ ____________

7. $\frac{40 \text{ ft}}{5 \text{ s}}$ ____________

Name ________________ Date ________ Class ________

LESSON 4-2

Puzzles, Twisters & Teasers

At This Rate...

Draw a line to connect each rate to the equivalent unit rate.

1. $\frac{\$22.80}{8\text{ lb}}$	$\frac{18\text{ ft}}{1\text{ s}}$	**O**
2. $\frac{288\text{ ft}}{16\text{ s}}$	$\frac{\$2.75}{1\text{ lb}}$	**I**
3. $\frac{\$16.50}{6\text{ lb}}$	$\frac{17.6\text{ ft}}{1\text{ s}}$	**M**
4. $\frac{792\text{ ft}}{45\text{ s}}$	$\frac{\$0.05}{1\text{ min}}$	**N**
5. $\frac{\$3.74}{68\text{ min}}$	$\frac{\$2.85}{1\text{ lb}}$	**K**
6. $\frac{\$4.05}{75\text{ min}}$	$\frac{\$0.06}{1\text{ min}}$	**A**

Next, choose the best buy. Circle the letter next to your answer.

7.	**E.** $\frac{\$3.33}{4.5\text{ lb}}$	**L.** $\frac{\$2.28}{3\text{ lb}}$	**C.** $\frac{\$2.92}{4\text{ lb}}$
8.	**O.** $\frac{\$45.36}{21\text{ gal}}$	**B.** $\frac{\$38.97}{18\text{ gal}}$	**T.** $\frac{\$57.25}{25\text{ gal}}$
9.	**S.** $\frac{\$2.16}{36\text{ min}}$	**A.** $\frac{\$3.30}{60\text{ min}}$	**G.** $\frac{\$5.04}{72\text{ min}}$
10.	**P.** $\frac{\$34.32}{8\text{ yd}}$	**W.** $\frac{\$49.92}{12\text{ yd}}$	**R.** $\frac{\$65.44}{16\text{ yd}}$

Write the letters that are next to your answers above the problem numbers to solve the riddle.

Where do tadpoles hang their coats?

___ ___ ___
3. 6. 5.

___ ___ ___ ___ ___ ___ ___ ___ ___
7. 10. 2. 9. 1. 10. 2. 8. 4.

Name ______________________ Date ______________ Class ______________

LESSON 4-3

Practice A

Identifying and Writing Proportions

Write the ratios in simplest form. Determine if the ratios are proportional by comparing them.

1. $\frac{1}{4}, \frac{3}{12}$ ______________

2. $\frac{2}{3}, \frac{6}{9}$ ______________

3. $\frac{4}{5}, \frac{15}{20}$ ______________

4. $\frac{3}{6}, \frac{6}{12}$ ______________

5. $\frac{5}{6}, \frac{16}{18}$ ______________

6. $\frac{2}{5}, \frac{6}{15}$ ______________

7. $\frac{1}{3}, \frac{3}{9}$ ______________

8. $\frac{4}{6}, \frac{7}{12}$ ______________

9. $\frac{3}{4}, \frac{18}{24}$ ______________

10. $\frac{2}{3}, \frac{9}{15}$ ______________

11. $\frac{2}{4}, \frac{9}{20}$ ______________

12. $\frac{3}{5}, \frac{15}{25}$ ______________

Find an equivalent ratio. Then write the proportion.

13. $\frac{1}{2}$ ______________

14. $\frac{3}{4}$ ______________

15. $\frac{5}{8}$ ______________

16. $\frac{4}{6}$ ______________

17. $\frac{1}{7}$ ______________

18. $\frac{10}{25}$ ______________

Name ____________________ Date ______________ Class ______________

LESSON 4-3

Practice B

Identifying and Writing Proportions

Determine whether the ratios are proportional.

1. $\frac{3}{4}, \frac{24}{32}$ ______________

2. $\frac{5}{6}, \frac{15}{18}$ ______________

3. $\frac{10}{12}, \frac{20}{32}$ ______________

4. $\frac{7}{10}, \frac{22}{30}$ ______________

5. $\frac{9}{6}, \frac{21}{14}$ ______________

6. $\frac{7}{9}, \frac{24}{27}$ ______________

7. $\frac{4}{10}, \frac{6}{15}$ ______________

8. $\frac{7}{12}, \frac{13}{20}$ ______________

9. $\frac{4}{9}, \frac{6}{12}$ ______________

10. $\frac{7}{8}, \frac{14}{16}$ ______________

11. $\frac{9}{10}, \frac{45}{50}$ ______________

12. $\frac{3}{7}, \frac{10}{21}$ ______________

Find a ratio equivalent to each ratio. Then use the ratios to write a proportion.

13. $\frac{7}{9}$ ______________

14. $\frac{11}{12}$ ______________

15. $\frac{14}{15}$ ______________

16. $\frac{35}{55}$ ______________

17. $\frac{14}{10}$ ______________

18. $\frac{25}{18}$ ______________

Name ______________________ Date ______________ Class ______________

LESSON 4-3

Practice C

Identifying and Writing Proportions

Determine whether the ratios are proportional.

1. $\frac{7}{11}, \frac{42}{60}$ ______________

2. $\frac{10}{18}, \frac{38}{72}$ ______________

3. $\frac{18}{28}, \frac{27}{42}$ ______________

4. $\frac{6}{14}, \frac{15}{35}$ ______________

5. $\frac{9}{24}, \frac{16}{40}$ ______________

6. $\frac{12}{39}, \frac{20}{65}$ ______________

Find a ratio equivalent to each ratio. Then use the ratios to write a proportion.

7. $\frac{7}{31}$ ______________

8. $\frac{24}{51}$ ______________

9. $\frac{6}{29}$ ______________

10. $\frac{14}{23}$ ______________

11. $\frac{17}{39}$ ______________

12. $\frac{25}{32}$ ______________

Complete each table of equivalent ratios.

13. 4 CDs to 10 tapes

CDs	2		10		28
Tapes		10		30	

14. 9 triangles per 6 circles

Triangles		9		30	
Circles	2		8		50

Find two ratios equivalent to each given ratio.

15. 10:21 ______________

16. 15:8 ______________

17. $\frac{5}{9}$ ______________

18. $\frac{24}{11}$ ______________

Name ______________________ Date ____________ Class ____________

LESSON 4-3

Review for Mastery

Identifying and Writing Proportions

Two ratios that are equal form a **proportion**. To determine whether two ratios are proportional, find the cross products of the ratios. If the cross products are equal, then the ratios are proportional.

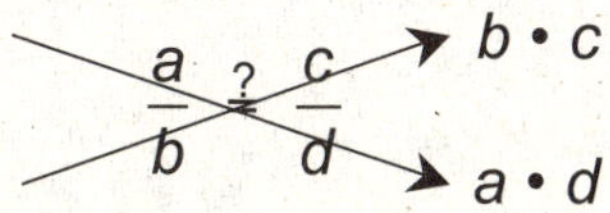

If $a \cdot d = b \cdot c$, then $\frac{a}{b} = \frac{c}{d}$.

Are $\frac{6}{9}$ and $\frac{8}{12}$ proportional?
Find the cross products.
$6 \cdot 12 = 72$ and $9 \cdot 8 = 72$
Since the cross products are equal, $\frac{6}{9}$ and $\frac{8}{12}$ are proportional.
So, $\frac{6}{9} = \frac{8}{12}$.

Are $\frac{4}{10}$ and $\frac{3}{8}$ proportional?
Find the cross products.
$4 \cdot 8 = 32$ and $10 \cdot 3 = 30$
Since the cross products are not equal, $\frac{4}{10}$ and $\frac{3}{8}$ are not proportional.
So, $\frac{4}{10} \neq \frac{3}{8}$.

Find the cross products to determine if the ratios are proportional.

1. $\frac{15}{21}, \frac{5}{7}$

   15 • ____ = ____ 21 • ____ = ____

   Are the ratios proportional? ______

2. $\frac{6}{9}, \frac{9}{15}$

   6 • ____ = ____ ____ • ____ = ____

   Are the ratios proportional? ______

3. $\frac{15}{9}, \frac{9}{4}$

   ____

4. $\frac{12}{24}, \frac{5}{10}$

   ____

5. $\frac{20}{12}, \frac{15}{9}$

   ____

You can write a proportion from a given ratio. Multiply or divide the numerator and denominator of the ratio by the same number.

$\frac{9}{12} = \frac{9 \div 3}{12 \div 3} = \frac{3}{4}$ So, $\frac{9}{12} = \frac{3}{4}$. $\frac{9}{12} = \frac{9 \cdot 4}{12 \cdot 4} = \frac{36}{48}$ So, $\frac{9}{12} = \frac{36}{48}$.

Find an equivalent ratio. Then write the proportion.

6. $\frac{6}{10}$

   ____

7. $\frac{10}{15}$

   ____

8. $\frac{18}{24}$

   ____

Name ______________________ Date ______________ Class ______________

LESSON 4-3

Challenge

What's in the Set?

The ratios below describe a set of polygons. Use the ratios to find the number of each type of polygon in the set. All ratios are written in simplest form.

- Equilateral triangles to isosceles triangles is 2 to 3
- Isosceles triangles to scalene triangles is 6:5
- Triangles to rectangles is $\frac{5}{4}$
- Parallelograms to triangles is $\frac{4}{3}$

1. What is the fewest number of each polygon that can be in the set?
   a. Equilateral triangles ______
   b. Isosceles triangles ______
   c. Scalene triangles ______
   d. All triangles ______
   e. Rectangles ______
   f. Parallelograms ______

2. Explain the order in which you determined the number of each polygon in the set.

____

____

____

3. Other sets of polygons with the same ratios are possible. Find the number of polygons in another set that has the same ratios.
   a. Equilateral triangles ______
   b. Isosceles triangles ______
   c. Scalene triangles ______
   d. All triangles ______
   e. Rectangles ______
   f. Parallelograms ______

Name ______________________ Date ______________ Class ______________

LESSON 4-3

Problem Solving

Identifying and Writing Proportions

Write the correct answer.

1. Jeremy earns $234 for 36 hours of work. Miguel earns $288 for 40 hours of work. Are the pay rates of these two people proportional? Explain.

____

2. Marnie bought two picture frames. One is 5 inches by 8 inches. The other is 15 inches by 24 inches. Are the ratios of length to width proportional for these frames? Explain.

____

3. The ratio of adults to children at a picnic is 4 to 5. The total number of people at the picnic is between 20 and 30. Write an equivalent ratio to find how many adults and children are at the picnic.

____

4. A recipe for fruit punch calls for 2 cups of pineapple juice for every 3 cups of orange juice. Write an equivalent ratio to find how many cups of pineapple juice should be used with 12 cups of orange juice.

____

Choose the letter for the best answer.

5. A clothing store stocks 5 blouses for every 3 pairs of pants. Which ratio is proportional for the number of pairs of pants to blouses?

   A 15:9
   C 12:20
   B 3:8
   D 18:25

6. To make lemonade, you can mix 4 teaspoons of lemonade powder with 16 ounces of water. What is the ratio of powder to water?

   F 4:32
   H 24:64
   G 32:8
   J 32:128

7. The town library is open 4 days per week. Suppose you use the ratio of days open to days in a week to find the number of days open in 5 weeks. What proportion could you write?

   A $\frac{4}{7} = \frac{20}{25}$
   C $\frac{4}{7} = \frac{20}{28}$
   B $\frac{7}{4} = \frac{21}{12}$
   D $\frac{4}{7} = \frac{20}{35}$

8. At a factory, the ratio of defective parts to total number of parts is 3:200. Which is an equivalent ratio?

   F 6:1000
   H 30:1000
   G 150:10,000
   J 1,000:10,000

Name ______________________ Date ______________ Class ______________

LESSON 4-3

Reading Strategies

Compare and Contrast

A **proportion** is two equal ratios.

Here are two ratios: $\frac{6}{8}$ and $\frac{9}{12}$.

To find out if they are equal, reduce ratios to simplest form.

$\frac{6}{8} = \frac{3}{4}$ $\frac{9}{12} = \frac{3}{4}$

$\frac{6}{8}$ and $\frac{9}{12}$ are **equal ratios**. They form a proportion.

Read: "6 is to 8 as 9 is to 12."

Compare these two ratios: $\frac{4}{7}$ and $\frac{5}{9}$.

These ratios are in simplest form, but they are *not equal*.

→ $\frac{4}{7} \neq \frac{5}{9}$

$\frac{4}{7}$ and $\frac{5}{9}$ are not equal ratios. They do *not* form a proportion.

Use the ratios $\frac{4}{6}$ and $\frac{1}{3}$ to answer Exercises 1–3.

1. Reduce $\frac{4}{6}$ to simplest form. ______________

2. Compare $\frac{4}{6}$ and $\frac{1}{3}$. Are they equal ratios?

____

3. Do these two ratios form a proportion? Why or why not?

____

Use the ratios $\frac{2}{5}$ and $\frac{4}{10}$ for Exercises 4–6.

4. Reduce $\frac{4}{10}$ to simplest form. ______________

5. Compare $\frac{2}{5}$ and $\frac{4}{10}$. Are they equal ratios?

____

6. Do $\frac{2}{5}$ and $\frac{4}{10}$ form a proportion? Why or why not?

____

Name ______________________ Date ______________ Class ______________

LESSON 4-3

Puzzles, Twisters & Teasers

Draw the Line!

Draw lines to connect the proportional ratios. Then start with number 1 and find the letter next to the answer. Use the letters with each correct answer to solve the riddle.

1.	$\frac{8}{14}$	$\frac{20}{16}$	**T**
2.	$\frac{2}{4}$	$\frac{3}{9}$	**E**
3.	$\frac{18}{24}$	$\frac{5}{6}$	**R**
4.	$\frac{10}{8}$	$\frac{2}{7}$	**S**
5.	$\frac{6}{21}$	$\frac{2}{10}$	**D**
6.	$\frac{6}{20}$	$\frac{3}{10}$	**Y**
7.	$\frac{1}{3}$	$\frac{25}{15}$	**A**
8.	$\frac{1}{5}$	$\frac{3}{4}$	**I**
9.	$\frac{5}{3}$	$\frac{12}{24}$	**H**
10.	$\frac{10}{12}$	$\frac{4}{7}$	**U**

Why do soccer players do well in math?

____ ____ ____ ____   ____ ____ ____
4 2 7 6   1 5 7

____ ____ ____ ____ ____   ____ ____ ____ ____ ____.
4 2 7 3 10   2 7 9 8 5

Name ______________________ Date ______________ Class ______________

LESSON 4-4

Practice A

Solving Proportions

Find the cross products.

1. $\frac{1}{2} = \frac{x}{8}$ ______

2. $\frac{a}{6} = \frac{7}{9}$ ______

3. $\frac{5}{b} = \frac{8}{10}$ ______

Use cross products to solve each proportion.

4. $\frac{2}{5} = \frac{x}{10}$ ______

5. $\frac{1}{3} = \frac{z}{15}$ ______

6. $\frac{3}{8} = \frac{s}{16}$ ______

7. $\frac{4}{r} = \frac{1}{4}$ ______

8. $\frac{10}{h} = \frac{5}{6}$ ______

9. $\frac{1}{d} = \frac{4}{12}$ ______

10. $\frac{w}{9} = \frac{6}{18}$ ______

11. $\frac{t}{8} = \frac{3}{4}$ ______

12. $\frac{k}{5} = \frac{9}{15}$ ______

13. $\frac{3}{6} = \frac{1}{f}$ ______

14. $\frac{2}{7} = \frac{6}{d}$ ______

15. $\frac{2}{9} = \frac{4}{c}$ ______

16. $\frac{a}{20} = \frac{15}{10}$ ______

17. $\frac{21}{k} = \frac{7}{4}$ ______

18. $\frac{3}{8} = \frac{n}{40}$ ______

19. Yolanda drove 50 miles in 2 hours at a constant speed. Use a proportion to find how long it would take her to drive 150 miles at the same speed.

____

Name ______________________ Date ______________ Class ______________

LESSON 4-4

Practice B

Solving Proportions

Use cross products to solve each proportion.

1. $\frac{2}{5} = \frac{x}{35}$ ______

2. $\frac{7}{r} = \frac{1}{4}$ ______

3. $\frac{k}{75} = \frac{9}{15}$ ______

4. $\frac{1}{3} = \frac{z}{27}$ ______

5. $\frac{2}{11} = \frac{12}{d}$ ______

6. $\frac{24}{s} = \frac{4}{12}$ ______

7. $\frac{w}{42} = \frac{6}{7}$ ______

8. $\frac{t}{54} = \frac{2}{9}$ ______

9. $\frac{3}{8} = \frac{a}{64}$ ______

10. $\frac{17}{34} = \frac{7}{f}$ ______

11. $\frac{15}{h} = \frac{5}{6}$ ______

12. $\frac{4}{15} = \frac{36}{c}$ ______

13. $\frac{z}{25} = \frac{12}{5}$ ______

14. $\frac{36}{k} = \frac{9}{4}$ ______

15. $\frac{5}{14} = \frac{n}{42}$ ______

16. $\frac{8}{9} = \frac{40}{m}$ ______

17. $\frac{7}{c} = \frac{63}{54}$ ______

18. $\frac{24}{21} = \frac{s}{35}$ ______

19. $\frac{e}{22} = \frac{6}{15}$ ______

20. $\frac{3}{v} = \frac{12}{17}$ ______

21. $\frac{5}{14} = \frac{4}{a}$ ______

22. Eight oranges cost $1.00. How much will 5 dozen oranges cost?

____

23. A recipe calls for 2 eggs to make 10 pancakes. How many eggs will you need to make 35 pancakes?

____

Name ______________________ Date ______________ Class ______________

LESSON 4-4

Practice C

Solving Proportions

Use cross products to solve each proportion.

1. $\frac{3}{7} = \frac{x}{49}$ ______

2. $\frac{4}{11} = \frac{z}{55}$ ______

3. $\frac{16}{9} = \frac{64}{a}$ ______

4. $\frac{13}{r} = \frac{1}{5}$ ______

5. $\frac{17}{41} = \frac{34}{f}$ ______

6. $\frac{k}{18} = \frac{11}{3}$ ______

7. $\frac{w}{39} = \frac{7}{13}$ ______

8. $\frac{7}{19} = \frac{t}{95}$ ______

9. $\frac{65}{j} = \frac{13}{17}$ ______

10. $\frac{15}{h} = \frac{5}{17}$ ______

11. $\frac{1.7}{3} = \frac{d}{21}$ ______

12. $\frac{5}{19} = \frac{35}{c}$ ______

13. $\frac{28}{9} = \frac{19.6}{m}$ ______

14. $\frac{e}{136} = \frac{13}{17}$ ______

15. $\frac{3.7}{3} = \frac{s}{21}$ ______

Arrange the four numbers to form a proportion that is true.

16. 40, 50, 80, 64 ______

17. 7, 9, 56, 72 ______

18. 50, 45, 15, 150 ______

19. A farmer can harvest 54 pounds of corn from each acre in his field. How much corn can he get from 12.5 acres?

______

20. A recipe for lasagna calls for 3 pounds of tomatoes to serve 5 people. A caterer wants to make enough lasagna to serve 110 people. How many pounds of tomatoes does he need?

______

Name ________________________ Date ______________ Class ______________

LESSON 4-4

Review for Mastery

Solving Proportions

Solving a proportion is like solving an equation involving fractions.

- Multiply both sides of the equation by the denominator of the fraction containing the variable.
- If the variable is in the denominator, invert both fractions in the proportion.

$$\frac{n}{7} = \frac{20}{28}$$

$$\cancel{7} \cdot \frac{n}{\cancel{7}} = 7 \cdot \frac{20}{28}$$

$$n = \frac{7 \cdot 20}{28} = \frac{140}{28}$$

$$n = 5$$

$$\frac{12}{x} = \frac{9}{6}$$

$$\frac{x}{12} = \frac{6}{9}$$

Invert both fractions.

$$\cancel{12} \cdot \frac{x}{\cancel{12}} = 12 \cdot \frac{6}{9}$$

$$x = \frac{12 \cdot 6}{9} = \frac{72}{9}$$

$$x = 8$$

Solve the proportion.

1. $\frac{a}{2} = \frac{27}{18}$

$\_\_\_ \cdot \frac{a}{2} = \frac{27}{18} \cdot \_\_\_$

$a = \frac{27 \cdot \_\_\_}{18}$

$a = \frac{\_\_\_}{18}$

$a = \_\_\_$

2. $\frac{8}{12} = \frac{n}{9}$

$\_\_\_ \cdot \frac{8}{12} = \frac{n}{9} \cdot \_\_\_$

$\frac{\_\_\_ \cdot 8}{12} = n$

$\frac{\_\_\_}{12} = n$

$n = \_\_\_$

3. $\frac{10}{t} = \frac{4}{6}$

$\frac{t}{10} = \frac{6}{4}$

$\_\_\_ \cdot \frac{t}{10} = \frac{6}{4} \cdot \_\_\_$

$t = \frac{6 \cdot \_\_\_}{4}$

$t = \frac{\_\_\_}{4}$

$t = \_\_\_$

4. $\frac{x}{15} = \frac{8}{10}$ ______________

5. $\frac{7}{3} = \frac{w}{18}$ ______________

6. $\frac{3}{2} = \frac{15}{c}$ ______________

Name ______________________ Date ______________ Class ______________

LESSON 4-4

Review for Mastery

Solving Proportions (continued)

You can use proportions to solve word problems.

A fruit punch is made with 32 ounces of ginger ale for every 12 ounces of frozen orange juice concentrate. How much ginger ale should you use for 30 ounces of orange juice concentrate?

- Set up a proportion comparing the amounts of ginger ale to orange juice concentrate.
- The first ratio shows the given recipe for the fruit punch.
- The second ratio shows the unknown amount of ginger ale as the variable *g*.
- Then solve the proportion.

1st ratio ↓ 2nd ratio ↓

$$\frac{\text{ginger ale}}{\text{orange juice concentrate}} = \frac{32}{12} = \frac{g}{30}$$

$$\frac{g}{30} = \frac{32}{12}$$

$$\cancel{30} \cdot \frac{g}{\cancel{30}} = 30 \cdot \frac{32}{12}$$

$$g = \frac{30 \cdot 32}{12} = \frac{960}{12}$$

$$g = 80$$

You should use 80 ounces of ginger ale for 30 ounces of frozen orange juice concentrate.

Solve.

7. Pecans cost $8.25 for 3 pounds. What is the cost of 5 pounds of pecans?

$\frac{\text{dollars}}{\text{pounds}} = \frac{\_\_\_\_}{\_\_\_\_} = \frac{c}{\_\_\_\_}$

$c =$ ______________

8. Mandy drove 90 miles in 2 hours at a constant speed. How long would it take her to drive 225 miles at the same speed?

$\frac{\text{miles}}{\text{hours}} = \frac{\_\_\_\_}{\_\_\_\_} = \frac{\_\_\_\_}{h}$

$h =$ ______________

9. Last week Geraldo bought 7 pounds of apples for $5.95. This week apples are the same price, and he buys 4 pounds. How much will he pay?

____

10. Aretha can type 55 words per minute. At that rate, how long will it take her to type a letter containing 935 words?

____

Name ______________________ Date ______________ Class ______________

LESSON 4-4

Challenge

Directly Speaking

Two quantities vary directly if they change in the same direction.

- If one quantity increases, then the other increases.
- If one quantity decreases, then the other decreases.

Here are some examples of direct variation.

- Amount bought and total cost
- Number of inches on a map and number of actual miles

If Farmer Jones plants 8 acres that produce 144 crates of melons, how many acres will produce 1,152 crates of melons?

Use the following proportion to solve.

$$\frac{\text{acres for 144 crates}}{\text{144 crates}} = \frac{\text{acres for 1,152 crates}}{\text{1,152 crates}}$$

$$\frac{8}{144} = \frac{x}{1{,}152} \qquad 144x = 9{,}216 \qquad x = 64$$

It will take 64 acres to produce 1,152 crates of melons.

Complete the proportion to find the answer.

		Do quantities increase or decrease?	Proportion	Answer
1.	If a furnace uses 40 gallons of oil in 8 days, how many gallons does it use in 10 days?		$\frac{\_\_\text{ gal}}{8\text{ d}} = \frac{x\text{ gal}}{\_\_\text{ d}}$	
2.	If 20 yards of wire weigh 80 pounds, what is the weight of 2 yards of the same wire?		$\frac{20\text{ yd}}{\_\_\text{ lb}} = \frac{\_\_}{x\text{ lb}}$	
3.	A cookie recipe calls for 2.5 cups of sugar and 4 eggs. If 6 eggs are used, how much sugar is needed?		$\frac{\_\_\ c}{4} = \frac{\_\_\ c}{\_\_}$	
4.	The scale on a map is 1 inch = 60 miles. What distance on the map represents 300 miles?		$\frac{\_\_}{\_\_} = \frac{x\text{ in.}}{300\text{ mi}}$	
5.	At the market, 4 limes cost $1.50. How much will 10 limes cost?		$\frac{\_\_}{\_\_} = \frac{10}{\_\_}$	

Name ______________________ Date ____________ Class ____________

LESSON 4-4

Problem Solving

Solving Proportions

Write the correct answer.

1. Euros are currency used in several European countries. On one day in October 2005, you could exchange \$3 for about 2.5 euros. How many dollars would you have needed to get 8 Euros?

____

2. A 3-ounce serving of tuna fish provides 24 grams of protein. How many grams of protein are in a 10-ounce serving of tuna fish?

____

3. Hooke's law states that the distance a spring is stretched is directly proportional to the force applied. If 20 pounds of force stretches a spring 4 inches, how much will the spring stretch if 80 pounds of force is applied?

____

4. Beeswax used in making candles is produced by honeybees. The honeybees produce 7 pounds of honey for each pound of wax they produce. How many pounds of honey is produced if 145 pounds of beeswax?

____

Choose the letter for the best answer.

5. For every 5 books her students read, Mrs. Fenway gives them a free homework pass for 4 days. Juan has accumulated homework passes for 12 days so far. What proportion would you write to find how many books Juan has read?

A $\frac{4}{12} = \frac{x}{5}$

B $\frac{4}{5} = \frac{x}{12}$

C $\frac{4}{5} = \frac{12}{x}$

D $\frac{5}{12} = \frac{4}{x}$

6. In his last 13 times at bat in the township baseball league, Santiago got 8 hits. If he is at bat 65 times for the season, how many hits will he get if his average stays the same?

F $\frac{8}{65} = \frac{x}{13}$

G $\frac{x}{65} = \frac{13}{8}$

H $\frac{8}{x} = \frac{65}{13}$

J $\frac{8}{13} = \frac{x}{65}$

7. A 12-pack of 8-ounce juice boxes costs \$5.40. How much would an 18-pack of juice boxes cost if it is proportionate in price?

A \$9.40 C \$3.60

B \$8.10 D \$12.15

8. Jeanette can swim 105 meters in 70 seconds. How far can she probably swim in 30 seconds?

F 20 meters H 45 meters

G 245 meters J 55 meters

Name ______________________ Date ______________ Class ______________

LESSON 4-4

Reading Strategies

Draw a Conclusion

There is a quick method to check to see if two ratios are equal. It is called the **cross products** method. Follow these steps.

Step 1: Multiply factors that cross diagonally in the two ratios.

$\frac{4}{5} \times \frac{12}{15}$ ← (5 × 12) ← (4 × 15)

Step 2: If the products of cross factors are the same, the two ratios are equal.

The cross products of 60 are the same.

So, $\frac{4}{5}$ and $\frac{12}{15}$ are equal ratios. → $\frac{4}{5} \times \frac{12}{15}$ (60, 60)

Two equal ratios form a **proportion.** $\frac{4}{5} = \frac{12}{15}$ is a proportion.

If cross products are not the same, the two ratios are not equal.

Is $\frac{4}{6} \stackrel{?}{=} \frac{5}{9}$?

5 × 6 = 30

$\frac{4}{6} \times \frac{5}{9}$

4 × 9 = 36

The cross products are not equal.

$\frac{4}{6} = \frac{5}{9}$ **is not** a proportion, so $\frac{4}{6} \neq \frac{5}{9}$.

Use the cross products method to tell whether the two ratios are equal. Show your work. Write yes or no.

1. $\frac{2}{4}$ and $\frac{6}{12}$ ____________

2. $\frac{2}{5}$ and $\frac{5}{10}$ ____________

3. $\frac{6}{9}$ and $\frac{8}{12}$ ____________

4. $\frac{3}{9}$ and $\frac{4}{10}$ ____________

5. How can you tell if two ratios form a proportion?

____

Name ________________ Date ____________ Class ____________

LESSON 4-4

Puzzles, Twisters & Teasers

The Proper Proportions!

Decide whether each pair of ratios is a proportion. Circle the letter above your answer. Then start with number 1, and use the letters to solve the riddle.

1. $\frac{2}{5} = \frac{6}{15}$

   A correct | J incorrect

2. $\frac{6}{10} = \frac{36}{60}$

   L correct | P incorrect

3. $\frac{4}{7} = \frac{5}{6}$

   K correct | L incorrect

4. $\frac{4}{8} = \frac{12}{24}$

   T correct | M incorrect

5. $\frac{1}{3} = \frac{86}{255}$

   N correct | H incorrect

6. $\frac{39}{4} = \frac{121}{12}$

   B correct | E incorrect

7. $\frac{2}{15} = \frac{12}{90}$

   F correct | V incorrect

8. $\frac{18}{90} = \frac{1}{5}$

   A correct | C incorrect

9. $\frac{45}{9} = \frac{15}{3}$

   N correct | Z incorrect

10. $\frac{34}{6} = \frac{96}{16}$

    L correct | S incorrect

11. $\frac{3}{24} = \frac{4}{52}$

    K correct | L incorrect

12. $\frac{14}{20} = \frac{5}{8}$

    G correct | E incorrect

13. $\frac{2}{5} = \frac{3}{12}$

    U correct | F incorrect

14. $\frac{35}{4} = \frac{175}{20}$

    T correct | E incorrect

Why did it get hot after the baseball game?

___ ___ ___

___ ___ ___

___ ___ ___ ___

___ ___ ___ ___.

Family Letter

4B Measurements

Dear Family,

The student is learning about customary and metric measures and unit conversion factors. Customary units of measure are commonly used in the United States. The table below lists benchmarks to help you understand some customary units.

	Customary Unit	Benchmark
Length	Inch (in.)	Length of a small paper clip
	Foot (ft)	Length of a standard sheet of paper
	Mile (mi)	Length of 4 laps around a track
Weight	Ounce (oz)	Weight of a slice of bread
	Pound (lb)	Weight of 3 apples
	Ton	Weight of a buffalo
Capacity	Fluid ounce (fl oz)	Amount of water in 2 tablespoons
	Cup (c)	Capacity of a measuring cup
	Gallon (gal)	Capacity of a large milk jug

The following table shows some common equivalent customary units. The student can use equivalent measures to convert from one unit of measure to another.

Length	Weight	Capacity
12 inches = 1 foot	16 ounces = 1 pound	8 fluid ounces = 1 cup
3 feet = 1 yard	2,000 pounds = 1 ton	2 cups = 1 pint
5,280 feet = 1 mile		2 pints = 1 quart
1,760 yards = 1 mile		4 quarts = 1 gallon

Convert 52 oz to pounds.

$$\frac{\text{ounces}}{\text{pounds}} \rightarrow \frac{16}{1} = \frac{52}{x}$$

Write a proportion using a ratio of equivalent measures.

$16x = 52$

Cross multiply.

$$\frac{16x}{16} = \frac{52}{16}$$

Divide both sides by 16.

$$x = \frac{52}{16} = 3\frac{4}{16} = 3\frac{1}{4}$$

Simplify.

52 ounces is equal to $3\frac{1}{4}$ pounds.

Vocabulary

These are the math words we are learning:

unit conversion factor a fraction in which the numerator and the denominator represent the same quantity in different units

Family Letter

4B Measurements *continued*

The metric system of measure is based on powers of 10. Metric units are used throughout the world. The table below lists benchmarks to help you understand some metric units.

	Metric Unit	Benchmark
Length	Millimeter (mm)	Thickness of a dime
	Centimeter (cm)	Width of your little finger
	Meter (m)	Width of a doorway
	Kilometer (km)	Length of 10 football fields
Mass	Milligram (mg)	Mass of a grain of sand
	Gram (g)	Mass of a small paperclip
	Kilogram (kg)	Mass of a textbook
Capacity	Milliliter (mL)	Capacity of an eyedropper
	Liter (L)	Capacity of a large water bottle
	Kiloliter (kL)	Capacity of 2 large refrigerators

The student will convert units within the metric system by multiplying or dividing by powers of 10. To convert to a smaller unit, you must multiply. To convert to a larger unit, you divide.

Convert 6.2 L to milliliters.

$6.2 \text{ L} = (6.2 \times 1{,}000) \text{ mL}$ Multiply by 1,000.

$= 6{,}200 \text{ mL}$ Move the decimal point 3 places to the right.

A **unit conversion factor** is a fraction in which the numerator and the denominator represent the same quantity in different units. The student will use a unit conversion factor to convert measurements from one system to another.

Convert 35 inches to centimeters. One inch is 2.54 centimeters.

$\frac{\text{inches}}{\text{centimeters}} \rightarrow \frac{1}{2.54} = \frac{35}{x}$ Write a proportion using 1 in. = 2.54 cm.

$1 \cdot x = 2.54 \cdot 35$ Cross multiply.

$x = 88.9$

35 inches is equal to 88.9 centimeters.

Sincerely,

Name ______________________ Date ______________ Class ______________

CHAPTER 4

At-Home Practice

4B Measurements

Choose the most appropriate customary unit for each measurement.

1. the length of a hallway
2. the capacity of a coffee mug
3. the weight of a car
4. the length of a calculator

Choose the most appropriate metric unit for each measurement.

5. the mass of a person
6. the capacity of a bathtub
7. the length of a swimming pool
8. the mass of an insect

Convert each measure.

9. 8 lb to ounces
10. 8.5 ft to inches
11. 21 c to pints
12. 13 gal to quarts
13. 155 mm to centimeters
14. 2.3 L to milliliters
15. 12.2 kg to grams
16. 19 km to meters

Convert between metric and customary units. Round to the nearest tenth.

17. 16 in. to centimeters. (Hint: 1 in. = 2.54 cm)
18. 5.5 lb. to kilograms (Hint: 1 lb ≈ 2.2 kg)
19. 24 mi to kilometers (Hint: 1 mi ≈ 1.609 km)
20. 31 L to quarts (Hint: 1 L ≈ 1.0567 qt)

Answers: 1. foot **2.** cup **3.** ton **4.** inch **5.** kilogram **6.** liter **7.** meter **8.** milligram **9.** 128 oz **10.** 102 in **11.** 10.5 pints **12.** 52 qt **13.** 15.5 cm **14.** 2,300 mL **15.** 12,200 g **16.** 19,000 m **17.** 40.6 cm **18.** about 12.1 kg **19.** about 38.6 km **20.** about 32.8 qt

Name ______________________ Date ______________ Class ______________

CHAPTER 4

Family Fun

Measure It!

You will need a ruler that has both inches and centimeters for this activity. Think of five objects in the household which are shorter than the ruler. These objects could be items such as a hairbrush, a book, or an envelope.

First estimate the lengths or widths of these items in both inches and centimeters. Then measure each item. The student will choose inches or centimeters to measure the items, and you will measure the items with the other unit.

When you have finished measuring, work together to convert the units: convert inches to centimeters and centimeters to inches. Remember that 1 inch is 2.54 centimeters.

Check whether your inch and centimeter measurements are similar to the conversions.

Finally, compare your measurements to your estimations. Were your estimations similar to the actual measurements?

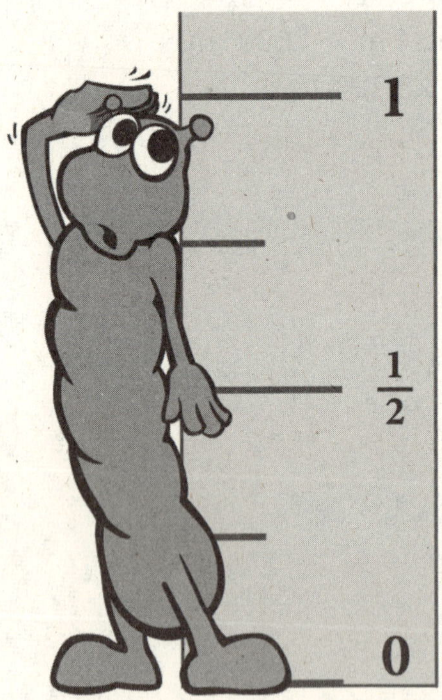

Name ______________________ Date ______________ Class ______________

LESSON 4-5

Practice A

Customary Measurements

Choose the letter of the best unit for each measurement.

1. The height of a house
   - A inches
   - B pounds
   - C feet
   - D miles

2. The weight of a letter
   - F ounces
   - G pounds
   - H inches
   - J tons

3. The capacity of a bathtub
   - A gallons
   - B cups
   - C fluid ounces
   - D pounds

4. The weight of a bowling ball
   - F ounces
   - G fluid ounces
   - H tons
   - J pounds

5. The thickness of a wallet
   - A ounces
   - B inches
   - C feet
   - D miles

6. The capacity of a spoon
   - F ounces
   - G fluid ounces
   - H cups
   - J gallons

Choose the letter of the best answer.

7. 8 gallons = ____ quarts
   - A 2
   - B 4
   - C 16
   - D 32

8. 96 ounces = ____ pounds
   - F 1536
   - G 16
   - H 8
   - J 6

9. 9 yards = ____ feet
   - A 3
   - B 12
   - C 27
   - D 108

10. 4.5 cups = ____ fluid ounces
    - F 2.25
    - G 9
    - H 36
    - J 72

11. 3 miles = ____ feet
    - A 15,840
    - B 10,560
    - C 1584
    - D 36

12. 16,000 pounds = ____ tons
    - F 4
    - G 8
    - H 80
    - J 1000

13. Awilda has 3 ft of ribbon.

    a. Write this measurement in inches.

    ____

    b. Suppose Awilda uses 20 in. of her ribbon to wrap a box. How much ribbon does she have left?

    ____________ in., or ____________ ft ____________ in.

Name ______________________ Date ______________ Class ______________

LESSON 4-5

Practice B

Customary Measurements

Choose the most appropriate customary unit for each measurement.
Justify your answer.

1. the weight of a paperback book

2. the capacity of a large soup pot

3. the length of a dining room table

4. the weight of an elephant

Convert each measure.

5. 6 mi to feet

6. 104 oz to pounds

7. 12 qt to pints

8. 5,000 lb to tons

9. 48 yd to feet

10. 96 fl oz to pints

11. 6.5 ft to inches

12. 20 qt to gallons

13. $3\frac{1}{4}$ lb to ounces

14. Marina has 2.5 lb of cashews. She puts 6 oz of cashews in a bag and gives the bag to her brother. What weight of cashews does Marina have left?

15. Faye is 5 ft 5 in. Faye is 10 in. shorter than her older brother. How tall is Faye's older brother?

Name ______________________ Date ______________ Class ______________

LESSON 4-5

Practice C

Customary Measurements

Choose the most appropriate customary unit for each measurement. Justify your answer.

1. the weight of an encyclopedia

____

____

2. the distance between two cities

____

____

3. the capacity of a medicine cup

____

____

4. the length of a hamster

____

____

Convert each measure.

5. 4.6 tons to pounds _______________

6. 6600 ft to miles _______________

7. 21 qt to pints _______________

8. 5 yd to inches _______________

9. 40 pt to gallons _______________

10. 148 oz to pounds _______________

Compare. Write <, >, or =.

11. 25,000 ft ____ 5 mi

12. 6 lb ____ 96 oz

13. 7.5 c ____ 56 fl oz

14. 4 ft ____ 32 in.

15. 2 gal ____ 48 c

16. 4 yd ____ 136 in.

17. Evan has 1 gal of orange juice. He uses 30 fl oz of orange juice to make a batch of smoothies. How much orange juice does Evan have left?

____

18. Helen has 3.5 yd of fabric. She buys 10 additional feet of fabric. How much fabric does Helen have now?

____

Name ________________________ Date ______________ Class ______________

LESSON 4-5

Review for Mastery

Customary Measurements

You can use the facts in this table to help you convert customary units.

Length	Weight	Capacity
12 in. = 1 ft	16 oz = 1 lb	8 fl oz = 1 c
3 ft = 1 yd	2,000 lb = 1 ton	2 c = 1 pt
5,280 ft = 1 mi		2 pt = 1 qt
		4 qt = 1 gal

To change larger units to smaller units, multiply.

3 lb = ■ oz

3 x 16 = 48

3 lb = 48 oz

1 lb = 16 oz
Multiply the number of pounds by 16.

To change smaller units to larger units, divide.

60 in. = ■ ft

60 ÷ 12 = 5

60 in. = 5 ft

12 in. = 1 ft
Divide the number of inches by 12.

Convert each measure.

1. 12 qt = ■ gal

   Smaller unit → larger unit

   Fact: _______ qt = 1 gal

   Operation: _______

   12 ÷ 4 = _______

   12 qt = ____ gal

2. 9 yd = ■ ft

   Larger unit → smaller unit

   Fact: 1 yd = _______ ft

   Operation: _______

   9 × _______ = _______

   9 yd = _______ ft

3. 7 lb = ■ oz

   _______ unit → _______ unit

   7 lb = _______ oz

4. 48 fl oz = ■ c

   _______ unit → _______ unit

   48 fl oz = _______ c

5. 8 ft = _______ in.

6. 4,000 lb = _______ tons

Name ______________________ Date ______________ Class ______________

LESSON 4-5

Challenge

Conversion Diversion

The table shows data about some of the most famous baseball parks in the United States. Use the table to answer the questions.

Baseball Field	Team	Year Opened	Year Closed	Cost	Dimensions of the Center Field Line
Ebbets Field	Brooklyn Dodgers	1913	1957	$750,000	393 ft
Forbes Field	Pittsburgh Pirates	1909	1970	$1 million	400 ft
Tiger Stadium	Detroit Tigers	1912	1999	$8 million	440 ft
Yankee Stadium	New York Yankees	1923		$50 million	408 ft
L.A. Coliseum	Los Angeles Dodgers	1958	1961	$950,000	420 ft

1. How many inches was the center field line in Tiger Stadium?

____

2. How many thousands of dollars did it cost to build Yankee Stadium?

____

3. For about how many months was the L.A. Coliseum open?

____

4. How many quarters (25¢) would it have taken to build the L.A. Coliseum?

____

5. How many inches longer was center field in Forbes Field than in Ebbets Field?

____

6. How many miles to the nearest hundredth of a mile was the center field line in the L.A. Coliseum?

____

7. For about how many months was Forbes Field open?

____

8. How many yards is the center field line in Yankee Stadium?

____

Name ______________________ Date ______________ Class ______________

LESSON 4-5

Problem Solving

Customary Measurements

Write the correct answer.

1. In 2003, a popcorn sculpture of King Kong was displayed in London. The sculpture was 13 ft tall and 8.75 ft wide. How many inches wide was the sculpture?

____

2. A pilot whale weighs 1500 lb. A walrus weighs 1.6 tons. Which weighs more? How much more?

____

3. A zoo has a rhesus monkey that weighed 20 lb. The monkey became sick and lost 18 oz. What was the monkey's new weight?

____

4. Two containers have capacities of 192 fl oz and 1.25 gal. Which container has a greater capacity? How much greater?

____

Choose the letter for the best answer.

This table gives lengths and weights for some apes.

Apes		
Name	**Maximum Height**	**Maximum Weight**
Chimpanzee	4 ft	115 lb
Gorilla	67.2 in.	0.2 tons
Orangutan	1.5 yd	3200 oz
Siamang	36 in.	240 oz

5. Which ape has the greatest weight?
   - A Chimpanzee
   - B Gorilla
   - C Orangutan
   - D Siamang

6. Which ape has the least height?
   - F Chimpanzee
   - G Gorilla
   - H Orangutan
   - J Siamang

7. Which ape has a maximum weight of 200 lb?
   - A Chimpanzee
   - B Gorilla
   - C Orangutan
   - D Siamang

8. Which two apes have a 6-inch difference in height?
   - F Chimpanzee and gorilla
   - G Gorilla and orangutan
   - H Siamang and chimpanzee
   - J Orangutan and chimpanzee

Name ______________________ Date ______________ Class ______________

LESSON 4-5

Reading Strategies

Use a Flowchart

This flowchart shows two ways to solve this problem: 56 oz = ■ pounds.

Identify equivalent measures that have the same units as the measures in the problem.
16 oz = 1 lb

↓

Choose a method to convert units.

Write a proportion.
Use a ratio of equivalent measures and a ratio of the measures in the problem.

ounces → $\frac{16}{1} = \frac{56}{x}$ ← pounds

↓

Solve to find the value of *x*.

$16 \cdot x = 56 \cdot 1$

$16x = 56$

$x = 3.5$

Write a multiplication equation.
Use a ratio of equivalent measures which equals 1. Set up the ratio so that you can cancel units.

$56 \text{ oz} = 56 \text{ oz} \times \frac{1 \text{ lb}}{16 \text{ oz}}$

↓

Simplify. $56 \text{ oz} = \frac{56}{1} \times \frac{1 \text{ lb}}{16 \text{ oz}}$

$= \frac{56 \text{ lb}}{16}$

$= 3.5 \text{ lb}$

56 oz = 3.5 pounds

Use the problem 18 yd = ■ ft for Exercises 1–6.

1. What equivalent measures can you use to solve this problem? ______________
2. What proportion can you write to solve this problem? __________
3. Solve the proportion you wrote. ________________
4. What multiplication equation can you write to solve this problem?

   ____

5. Simplify the equation you wrote. ________________
6. How do your answers to Exercises 3 and 5 compare?

   ____

Name ______________________ Date ______________ Class ______________

LESSON 4-5

Puzzles, Twisters & Teasers

Because It's Customary

Choose the most appropriate customary unit for each measurement. Circle the letter above your answer.

1. the capacity of a hot water heater

B	T
fluid ounces	gallons

2. the weight of a bowling ball

D	K
pounds	tons

3. the thickness of a book

N	C
inches	feet

4. the capacity of a thermos

A	H
gallons	quarts

Next, choose the measurement that is equivalent to the one given. Circle the letter above your answer.

5. 6 ft = ?

S	W
48 in.	72 in.

6. 12 oz = ?

C	O
0.75 lb	192 lb

7. 80 fl oz = ?

E	C
10 c	5 lb

8. 3.5 tons = ?

P	O
56 lb	7000 lb

9. 36 ft = ?

K	O
12 yd	3 yd

10. 40 qt = ?

I	Y
160 gal	10 gal

Write the circled letters above the problem numbers to solve the riddle.

Why do lions eat raw meat?

___ ___ ___ ___ ___ ___ ___ ___ ___ ___ ___ ___
1. 4. 7. 10. 2. 8. 3. 1. 9. 3. 8. 5.

___ ___ ___ ___ ___ ___ ___ ___ ___
4. 8. 5. 1. 8. 6. 8. 8. 9.

Name ______________________ Date ______________ Class ______________

LESSON 4-6

Practice A

Metric Measurements

Choose the letter for the best answer.

1. The height of a flagpole
   A millimeters C meters
   B centimeters D kilometers

2. The mass of a bicycle
   F milligrams H grams
   G kilograms J centimeters

3. The capacity of a spoon
   A liters C kiloliters
   B milliliters D millimeters

4. The thickness of a piece of cardboard
   F millimeters H kilometers
   G centimeters J meters

5. The mass of a banana
   A kilograms C liters
   B grams D milligrams

6. The capacity of a swimming pool
   F kilograms H liters
   G milliliters J kiloliters

Choose the letter of the best answer.

7. 320 cm = ____ m
   A 0.032 C 3.2
   B 0.32 D 32

8. 2.5 kg = ____ g
   F 0.0025 H 250
   G 0.25 J 2,500

9. 5,600 mL = ____ L
   A 5,600,000 C 56
   B 56,000 D 5.6

10. 780 m = ____ km
   F 780,000 H 0.78
   G 78,000 J 0.078

11. 0.35 g = ____ mg
   A 0.00035 C 350
   B 0.0035 D 3,500

12. 4.3 mm = ____ cm
   F 0.043 H 430
   G 0.43 J 4,300

13. Benjamin is 165 cm tall. His sister is 1.8 m tall. Who is shorter? Explain your answer.

    ____

14. Jan has a bag of grapes that has a mass of 0.9 kg. Terri has a bag of grapes that has a mass of 830 g. Whose grapes have the greater mass? Explain your answer.

    ____

Name ______________________ Date ______________ Class ______________

LESSON 4-6

Practice B

Metric Measurements

Choose the most appropriate metric unit for each measurement. Justify your answer.

1. The capacity of a paper cup

2. The mass of a small poodle.

3. The width of a computer screen

4. The mass of a pencil

Convert each measure.

5. 496 mm to centimeters

6. 0.68 kg to grams

7. 3,800 mL to liters

8. 832 mg to grams

9. 76 km to meters

10. 2.9 cm to meters

11. 0.041 kL to liters

12. 14.9 g to milligrams

13. 7,800 cm to meters

14. Sam's laptop computer has a mass of 4.2 kg. Fred's laptop computer has a mass of 4,940 grams. Which computer has the lesser mass? Explain your answer.

15. Elise makes a poster that is 1.5 m tall. Meg makes a poster that is 96 cm tall. Who makes a taller poster? Explain your answer.

Name ______________________ Date ______________ Class ______________

LESSON 4-6

Practice C

Metric Measurements

Choose the most appropriate metric unit for each measurement. Justify your answer.

1. The thickness of a piece of spaghetti

2. The capacity of a small sink

3. The mass of an envelope

4. The capacity of a gas truck

Convert each measure.

5. 3.06 kL to liters ______

6. 75 mm to centimeters ______

7. 604 g to kilograms ______

8. 0.95 m to centimeters ______

9. 5.6 L to milliliters ______

10. 48 mg to grams ______

Compare. Write >, <, or = for ____.

11. 0.5 cm ____ 5 mm

12. 4.6 kg ____ 462 g

13. 0.03 kL ____ 200 L

14. 53 m ____ 5,400 mm

15. 8,840 mL ____ 9 L

16. 416 mg ____ 0.4 g

17. Eleanor ran 3.5 km on Friday and 2.8 km on Saturday. Monroe ran 6,400 m on the same two days. Who runs a greater distance? Explain.

____

18. Tunnel A is 1.6 km long. Tunnel B is 740 meters long. Which tunnel is longer? How much longer?

____

Name ______________________ Date ______________ Class ______________

LESSON 4-6

Review for Mastery

Metric Measurements

You can use the relationships in this table to help you convert metric units.

Length	Mass	Capacity
1 cm = 10 mm	1 g = 1,000 mg	1 L = 1,000 mL
1 m = 100 cm = 1,000 mm	1 kg = 1,000 g	1 kL = 1,000 L
1 km = 1,000 m		

To change larger units to smaller units, multiply.

5.3 L = ____ mL
5.3 × 1,000 = 5,300
5.3 L = 5,300 mL

1 L = 1,000 mL
Liters are larger than milliliters, so multiply by 1,000.
Move the decimal point 3 places to the right: 5.300

To change smaller units to larger units, divide.

46 cm = ____ m
46 ÷ 100 = 0.46
46 cm = 0.46 m

1 m = 100 cm
Centimeters are smaller than meters, so divide by 100.
Move the decimal point 2 places to the left: 046

Convert each measure.

1. 15 cm to millimeters

   Think: 1 cm = ______________ mm

   Multiply 15 by ______________.

   15 cm = ______________ mm

2. 964 g to kilograms

   Think 1 kg = ______________ g

   Divide 964 by ______________

   964 g = ______________ kg

3. 4.9 kL to liters

   ______________ 4.9 by ______________

   4.9 kL = ______________ L

4. 3,531 cm to meters

   ______________ 3,531 by ______________

   3,531 cm = ______________ m

5. 468 mg to grams

   468 mg = ______________ g

6. 0.69 kL to liters

   0.69 kL = ______________ L

Name ______________________ Date ______________ Class ______________

LESSON 4-6

Challenge

Have a Ball!

The most commonly used metric units of mass are **milligrams (mg)**, **grams (g)**, and **kilograms (kg)**. However, there are other metric units of mass also. These units include **centigrams (cg)**, **decigrams (dg)**, **decagrams (dag)**, and **hectograms (hg)**. The table below shows the equivalent number of grams for each of these units.

Unit	**1 kg**	**1 hg**	**1 dag**	**1 g**	**1 dg**	**1 cg**	**1 mg**
Number of Grams	1,000 g	100 g	10 g	1 g	0.1 g	0.01 g	0.001 g

The table below shows the masses of balls that are used in different sports. Rewrite the table so that it shows the balls and their masses in order from greatest mass to least mass. Use the same unit to show each mass.

Ball	Mass
Baseball	14.88 dag
Basketball	0.65 kg
Football	4,252 dg
Paddleball	6,502 cg
Ping pong ball	2,450 mg
Softball	198.4 g
Soccer ball	4.536 hg
Tennis Ball	750 dg
Volleyball	28,000 cg

Ball	Mass

Name ______________________ Date ______________ Class ______________

LESSON 4-6

Problem Soving

Metric Measurements

Write the correct answer.

1. A porcupine has a mass of 27 kg. A cat has a mass of 6,300 g. Which animal has the greater mass?

____

2. A faucet drips at the rate of 50 L per day. At this rate, how many days will it take for the faucet to drip a total of 1 kL?

____

3. The distance from Midwood Library to Midwood Middle School is 0.845 km. The distance from Midwood High School to Midwood Library is 872 m. Which school is closer to the library? How much closer?

____

____

4. A mouse has a mass of 18 g. A guinea pig has a mass of 0.85 kg. Is the mass of the two animals together greater or less than 1 kg? Explain your answer.

____

____

Choose the letter for the best answer.

The table shows the lengths and masses of some members of the cat family.

Members of the Cat Family		
Name	**Length**	**Mass**
Bobcat	864 mm	18,200 g
Jaguar	1.52 m	55 kg
Canada Lynx	91.4 cm	15,900 g
Mountain Lion	1.35 m	85 kg

5. Which cat has the greatest length?

A bobcat

B jaguar

C Canada lynx

D mountain lion

6. Which cat has the least mass?

F bobcat

G jaguar

H Canada lynx

J mountain lion

7. What is the difference in mass between the cat with the greatest mass and the cat with the least mass?

A 66.8 kg

B 69.1 kg

C 7,400 g

D 15,815 g

8. For which two cats is the difference in length closest to 0.5 m?

F bobcat and mountain lion

G bobcat and Canada lynx

H jaguar and mountain lion

J jaguar and Canada lynx

Name ______________________ Date ______________ Class ______________

LESSON 4-6

Reading Strategies

Follow a Procedure

To convert metric units, follow these steps.

653 mg = ____ g

Step 1: Determine if you are converting to a larger unit or a smaller unit.

1 mg ≤ 1 g
You are converting to a larger unit.

Step 2: Choose the operation that you need to use to convert.

- To convert to a smaller unit, you need to multiply.
- To convert to a larger unit, you need to divide.

Since you are converting to a larger unit, you need to divide.

Step 3: Identify the number by which you must multiply or divide. That number is the number of smaller units that equals 1 larger unit.

1,000 mg = 1 g
So, you need to divide 653 mg by 1,000 to get the number of grams.

Step 4: Calculate the answer.

653 ÷ 1,000 = 0.653
653 mg = 0.653 g

Answer these questions for 3.9 m = ____ cm.

1. Are you converting to a larger unit or a smaller unit? ______________
2. What operation will you use to convert? ______________
3. By what number do you need to multiply or divide? ______________
4. What is the answer to the problem? ______________

Answer these questions for 14.8 L = ____ kL.

5. Are you converting to a larger unit or a smaller unit? ______________
6. What operation will you use to convert? ______________
7. By what number do you need to multiply or divide? ______________
8. What is the answer to the problem? ______________

Name ______________________ Date ______________ Class ______________

LESSON 4-6

Puzzles, Twisters & Teasers

Measure for Measure

Choose the most appropriate metric unit for each measurement. Circle the letter above your answer.

1. The capacity of a juice can

Y	T
milliliters	kiloliters

2. the length of a swimming pool

H	O
millimeters	meters

3. The mass of a bowling ball

E	U
milligrams	kilograms

4. The capacity of a walk-in freezer

R	L
kiloliters	milliliters

5. The width of a postcard

E	I
centimeters	kilometers

6. The mass of a button

T	A
kilograms	grams

Choose the measurement that is equivalent to the one given. Circle the letter above your answer.

7. 680 mg = ?

T	R
0.068 g	0.68 g

8. 13.7 cm = ?

D	L
137 mm	13.7 m

9. 0.49 kg = ?

E	R
4,900 g	490 g

10. 27.1 L = ?

U	T
0.0271 kL	271 mL

11. 964 mm = ?

M	O
0.964 m	9.64 cm

12. 0.78 m = ?

T	S
78 mm	78 cm

Start with problem number 1, and use the circled letters to solve the riddle.

What part of your body has the most rhythm?

____ ____ ____ ____

____ ____ ____ ____ ____ ____ ____ ____

Name ______________________ Date ______________ Class ______________

LESSON 4-7

Practice A

Dimensional Analysis

1. Bethany's dog eats 450 grams of food per day. Find this rate in kilograms per week.

____

2. Grace runs 3 miles a day. Find this rate in feet per day.

____

3. Jefferson drinks 10 cups of orange juice a day. Find this rate in pints per week.

____

4. One meter is about 3.28 feet. A door is 2.7 meters tall. What is the height of the door in feet, rounded to the nearest tenth of a foot?

____

5. One gallon is about 3.79 liters. A water cooler holds 40 liters. What is the capacity of the water cooler in gallons, rounded to the nearest tenth of gallon?

____

6. One pint is about 0.47 liters. A recipe calls for 7 pints of whipping cream. How much whipping cream does the recipe require in liters, rounded to the nearest tenth of a liter?

____

7. One pound is about 0.45 kilograms. Julie's bowling ball weighs 14 pounds. What is the weight of Julie's bowling ball, to the nearest tenth of a kilogram?

____

8. Aurora wants to improve her running speed. She can run 225 meters in 1.5 minutes. Find Aurora's speed in meters per minute. Use dimensional analysis to check the reasonableness of your answer.

____

Name ______________________ Date ______________ Class ______________

LESSON 4-7

Practice B

Dimensional Analysis

1. David takes 300 milligrams of medicine every day. Find this rate in grams per day.

____

2. Jody runs the 500-yard dash for his school's track team in 1 minute. Find this rate in yards per second.

____

3. Sean drinks six 12-fluid ounce cans of soda a week. Find this rate in pints per week.

____

4. One quart is about 0.946 liters. A recipe for punch requires diluting the punch concentrate with 7 quarts of water. To the nearest tenth, how many liters of water are required to dilute the concentrate according to the directions?

____

5. One pound is about 0.454 kilograms. Jesse's dog Angel weighs $18\frac{1}{2}$ pounds. To the nearest tenth, how many kilograms does Angel weigh?

____

6. One mile is about 1.61 kilometers. The volcano Mauna Kea is 120 km wide. What is the width of the volcano in miles, rounded to the nearest tenth of a mile?

____

7. A 1-pound weight has a mass of about 0.45 kilograms. A car weighs 2100 pounds. What is the mass of the car in kilograms, rounded to the nearest tenth of a kilogram?

____

8. Tyler is riding his bike from his house to the store, which is 4 miles away. Tyler gets to the store 18 minutes after he leaves his house. Find Tyler's speed in miles per minute. Use dimensional analysis to check the reasonableness of your answer.

____

Name ______________________ Date ______________ Class ______________

LESSON 4-7

Practice C

Dimensional Analysis

1. Barry's remote control car can move 5.5 feet in one second. Find this rate in miles per hour.

____

2. Karina's pool can drain 1.75 quarts per second. Find this rate in gallons per hour.

____

3. Some cheetahs can run as fast as 71 miles per hour. To the nearest hundredth, find this rate in feet per second.

____

4. The giant clam is a sea creature that can weigh up to 227 kg. To the nearest whole number, what is the weight in ounces of the giant clam? (Hint: 2.2046 lb $\approx$ 1 kg)

____

5. A 1-ounce weight has a mass of about 28.35 grams. An empty ostrich egg has a mass of 300 g. How much does it weigh in ounces, rounded to the nearest ounce?

____

6. An inch is about 2.54 centimeters. A plant measures 14.875 inches tall. What is the height of the plant in centimeters, rounded to the nearest tenth of a centimeter?

____

7. One yard is about 0.914 meters. The distance between first and second base on a baseball diamond is 30 yards. What is the distance between bases in meters, rounded to the nearest tenth of a meter?

____

8. Kara is jogging on a track that is 1500 ft. She can do 1 lap around the track in 4 minutes. Find Kara's speed in feet per minute. Use dimensional analysis to check the reasonableness of your answer.

____

Name ______________________ Date ______________ Class ______________

LESSON 4-7

Review for Mastery

Dimensional Analysis

To change from one unit to another, multiply by a fraction of value 1.

Convert 7 ft to in.

Use 1 foot = 12 inches to write a fraction of value 1.

Put the unit to be changed in the denominator. $\frac{12\text{ in.}}{1\text{ ft}}$

$7\,\cancel{\text{ft}} \times \frac{12\text{ in.}}{1\,\cancel{\text{ft}}} = (7 \times 12)\text{ in.} = 84\text{ in.}$

When converting from a larger unit to a smaller, the result is a larger number.

7 ft = 84 in.

When converting from a smaller unit to a larger, the result is a smaller number.

84 in. = 7 ft

Complete to convert each unit.

1. 96 inches to feet

   1 ft = __________

   unit for denominator of fraction is

   ____

   96 in. × —— = __________ ft

2. 56 ounces to pounds

   1 lb = __________

   unit for denominator of fraction is

   ____

   56 oz × —— = __________ lb

3. 635 meters to centimeters

   1 m = __________

   unit for denominator of fraction is

   ____

   635 m × —— = __________ cm

4. 172 pounds to kilograms

   1 kg = 2.2 lb

   unit for denominator of fraction is

   ____

   172 lb × —— ≈ __________ kg

5. 105 feet to yards

   1 yd = __________

   unit for denominator of fraction is

   ____

   105 ft × —— = __________ yd

6. 4375 milliliters to liters

   1L = __________

   unit denominator for fraction is

   ____

   4375 mL × —— = __________ L

Name ______________________ Date ______________ Class ______________

LESSON 4-7

Challenge

Water, Water, Everywhere

When backpacking it's important to carry enough supplies so that you can keep hydrated. Sometimes, that means including a supply of fresh water. When you have to carry your water with you, it's important to know how much water weighs. Answer the questions below to find the weight of a gallon of water.

1. The density of water is 1 gram per milliliter: $\frac{1\text{ g}}{1\text{ mL}}$.

   Use 1 mL = 1 cm^3 to find this rate in terms of grams per cubic centimeter.

   density of water = __________ g/cm^3

2. Since 1 m = 100 cm, how many cubic centimeters are there in 1 cubic meter?

   1 m^3 = __________ cm^3

3 The rate in Exercise 1 is expressed in grams per cubic centimeter. Change this rate to kilograms per cubic meter. Round your answer to the nearest hundredth.

   density of water = __________ kg/m^3

4. The rate in Exercise 3 is expressed in kilograms per cubic meter. Use 1 kg = 2.205 lb and 1 m^3 = 264.2 gal to change this rate to pounds per gallon.

   density of water = __________ lb/gal

5. So, one gallon of water weighs about __________ pounds.

Name ______________________ Date ______________ Class ______________

LESSON 4-7

Problem Solving

Dimensional Analysis

Use the following: 1 mile = 1.609 km; 1 kg = 2.2046 lb. Round to the nearest tenth.

1. Chinook salmon travel up to 3,000 km to find spawning grounds. How far will a Chinook salmon travel in miles?

____

2. Chinook salmon can travel at 23 km/h. How fast can the Chinook salmon travel in miles per hour?

____

3. The average rhinoceros weighs 1600 kg. How many pounds does the average rhinoceros weigh?

____

4. An adult rhinoceros can have a horn that weighs 3.5 kg. How many pounds do these horns weigh?

____

Paraceratherium was the biggest land mammal there has ever been. It lived about 35 million years ago and was 11 m long. The blue whale is the largest mammal that has ever lived. It can be as long as 33 m. 1 foot = 0.3048 meters. Round to the nearest tenth.

5. How long is the blue whale in feet?

____

6. How long was the paraceratherium in feet?

____

Round to the nearest tenth. Choose the letter for the best answer.

7. The world's fastest bird is the peregrine falcon that flies up to 200 mi an hour. How fast is the peregrine falcon in feet each second?

   A 136.4 ft/s C 543.2 ft/s
   B 293.3 ft/s D 9,680 ft/s

8. The longest gloved fight between two Americans lasted for more than seven hours before being declared a draw. How many seconds did the fight last?

   F 127 s H 420 s
   G 385 s J 25,200 s

9. The average person showers for eight minutes. How many seconds does it take the average person to shower?

   A 127 s C 480 s
   B 385 s D 28,800 s

10. The brain of an average adult female weighs 49 oz. How many pounds does the average female brain weigh?

   F 3.1 lb H 13.8 lb
   G 5.8 lb J 784 lb

Name ______________________ Date ____________ Class ____________

LESSON 4-7

Reading Strategies

Choose Relevant Information

Here are some common measurement units and their equivalents.

- 16 ounces in 1 pound
- 8 ounces in 1 cup
- 12 inches in 1 foot
- 60 minutes in 1 hour
- 4 quarts in 1 gallon
- 3 feet in 1 yard

You can change a measurement from one unit to another if you know the equivalent measure.

If a larger unit, such as feet, is changed into a smaller one, such as inches, you will have a greater number of units. If a smaller unit, such as quarts, is changed into a larger unit, such as gallons, you will have a smaller number of units.

1. 40 ounces = ___ pounds. Will the answer be larger or smaller than 40?

____

2. 21 feet = ___ yards. Will the answer be larger or smaller than 21?

____

3. 20 gallons = ___ quarts. Will the answer be larger or smaller than 20?

____

Answer the following questions to change 8 feet into inches.

4. Will the numeric portion of your answer be smaller or larger? Explain.

____

____

5. How many inches are in a foot?

____

6. How many feet are being changed into inches?

____

Multiply the numbers in Exercises 5 and 6 together to get the number of inches in 8 feet

7. How many inches are in 8 feet?

____

Name ______________________ Date ______________ Class ______________

LESSON 4-7

Puzzles, Twisters & Teasers

Analyze This!

Solve the crossword puzzle.

Across

4. Unit analysis is also called ___ analysis.
5. There are 100 centimeters in 1 ___.
6. There are 7 days in 1 ___.
8. There are 60 ___ in 1 minute.
10. There are 16 ___ in 1 pound.

Down

1. Ratios of equal quantities are called ___ factors.
2. There are 4 quarts in 1 ___.
3. There are 1000 meters in a ___.
7. There are 12 ___ in 1 foot.
9. There are 365 days in 1 ___.

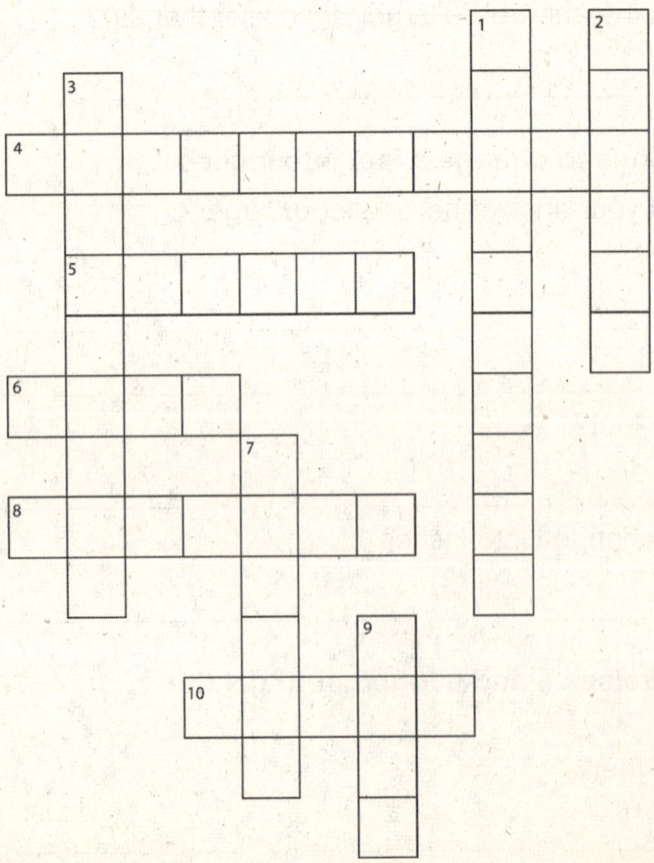

CHAPTER 4

Family Letter

4C Proportions in Geometry

Dear Family,

The student will learn to use ratios to determine whether two figures are **similar**. Similar figures have the same shape but not necessarily the same size.

To determine whether two figures are similar, you must compare the measures of the corresponding angles and the ratios of the lengths of the corresponding sides.

Similar Figures
Two figures are similar if • the measures of their corresponding angles are equal. • the ratios of the lengths of their corresponding sides are proportional.

Tell whether the triangles are similar.

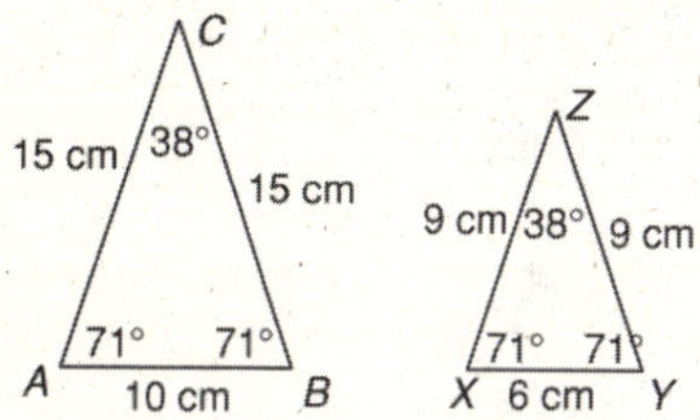

The corresponding angles of the figures have equal measures.

$\overline{AB}$ corresponds to $\overline{XY}$.

$\overline{AC}$ corresponds to $\overline{XZ}$.

$\overline{BC}$ corresponds to $\overline{YZ}$.

$\frac{AB}{XY} \stackrel{?}{=} \frac{AC}{XZ} \stackrel{?}{=} \frac{BC}{YZ}$ Write ratios using the corresponding sides.

$\frac{10}{6} \stackrel{?}{=} \frac{15}{9} \stackrel{?}{=} \frac{15}{9}$ Substitute the lengths of the sides.

$\frac{5}{3} = \frac{5}{3} = \frac{5}{3}$ Simplify each ratio.

Since the measures of the corresponding angles are equal and the ratios of the corresponding sides are equivalent, the triangles are similar.

Vocabulary

These are the math words we are learning:

corresponding angles the angles that are in the same relative position in two similar figures

corresponding sides the sides that are in the same relative position in two similar figures

indirect measurement a method of using proportions to find an unknown length or distance in similar figures

scale a ratio between two sets of measurements

scale drawing a proportional two-dimensional drawing of an object

scale factor a ratio that compares the size of a model to the actual size of the object

scale model a proportional three-dimensional model of an object

similar figures that have the same shape but not necessarily the same size

CHAPTER 4

Family Letter

4C Proportions in Geometry *continued*

The student will use the properties of similar figures to find the measure of unknown lengths or unknown distances by setting up proportions to find the missing value. This form of indirect measurement is applicable to many everyday situations.

Harry and Sally both have rectangular swimming pools that are similar to each other. How wide is the pool at Sally's house?

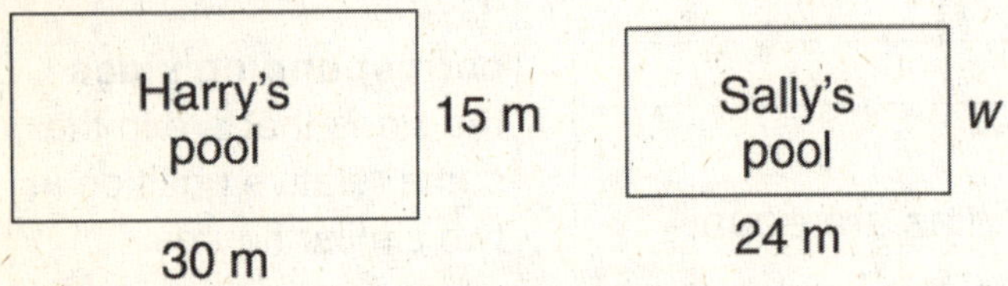

Let w = the width of Sally's pool.

$\frac{24}{30} = \frac{w}{15}$ Write a proportion using corresponding sides.

$30 \cdot w = 24 \cdot 15$ Find the cross products.

$30w = 360$ Multiply.

$\frac{30w}{30} = \frac{360}{30}$ Divide each side by 30.

$w = 12$

Sally's pool is 12 m wide.

Scale models and drawings are proportional representations of objects. The scale factor tells you how the dimensions of a model or drawing are related to the dimensions of the actual object. A scale factor is always the ratio of the model's dimensions to the actual object's dimensions.

An architect's model of a building is 18 inches tall. The actual building is 360 inches tall. What is the scale factor?

$\frac{\text{model height}}{\text{building height}} = \frac{18}{360} = \frac{1}{20}$ Write a ratio. Then simplify.

The scale factor is $\frac{1}{20}$.

The concepts and skills learned in this section are applicable to many activities outside of the math classroom. Provide opportunities for the student to use the skills learned in this chapter in real-world applications.

Sincerely,

Name ______________________ Date ______________ Class ______________

CHAPTER 4

At-Home Practice

4C Proportions in Geometry

Tell whether the triangles are similar.

1.

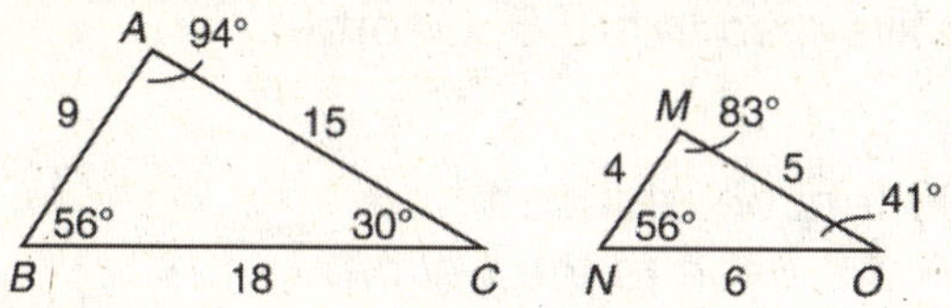

____

2.

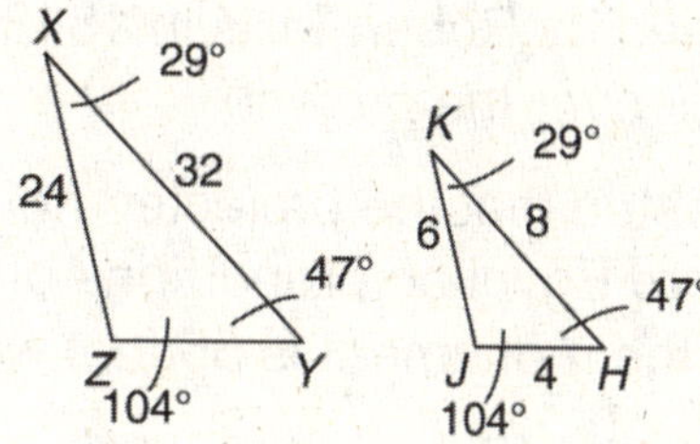

____

Tell whether the figures are similar.

3.

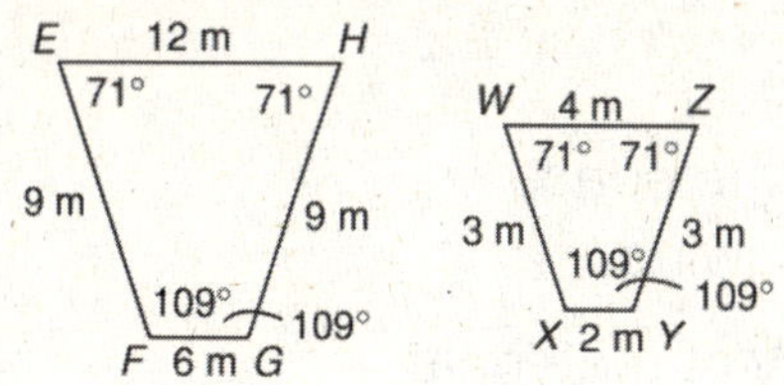

____

4.

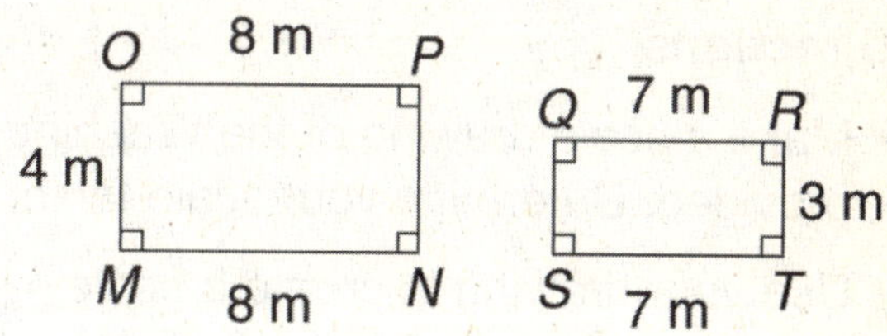

____

Find the unknown length in each pair of similar figures.

5.

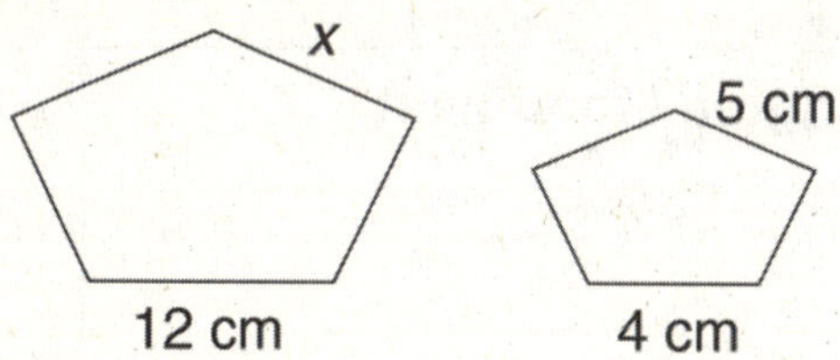

____

6.

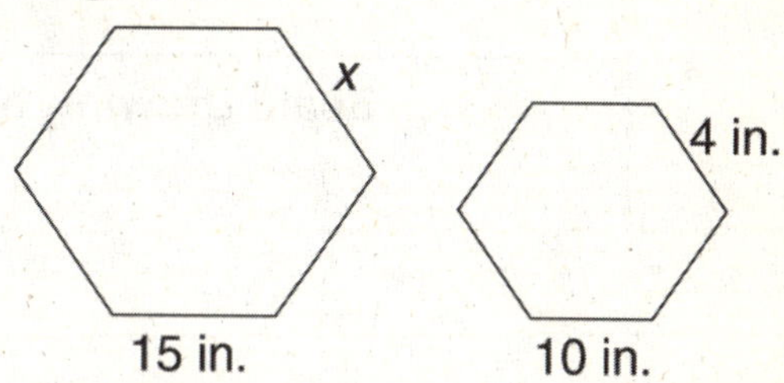

____

Identify the scale factor.

7.

	Car	Model
length (in.)	102	6

Scale Factor = ______________________

Answers: **1.** not similar **2.** similar **3.** similar **4.** not similar **5.** $x = 15$ cm **6.** $x = 6$ in. **7.** $\frac{1}{17}$

Name ______________________ Date ______________ Class ______________

CHAPTER 4

Family Fun

Scale It Down

The Washington Monument is the tallest all-masonry tower in the world. It was designed by Robert Mills in 1848. When Mills died in 1855, George P. Marsh took over the project.

Marsh suggested that the projected height of 600 feet be reduced to the standard Egyptian proportions of *ten times the base to the height.* The base of the monument is 55 feet wide.

What is the height? ______________________.

Finally, Marsh proposed that a pyramid cap of aluminum top this great monument. The project was completed in 1885.

Directions

- Make a scale drawing of the Washington monument in the space provided. Determine your scale factor and label your drawing.
- Use your drawing to create a scale model. Use a variety of materials such as modeling clay, craft sticks, milk cartons, blocks, etc.
- Investigate other famous monuments or structures in order to create another scale model.

Scale Drawing of the Washington Monument

Scale Factor __________

Answer: 550 ft

Name ______________________ Date ______________ Class ______________

LESSON 4-8

Practice A

Similar Figures and Proportions

Identify the corresponding sides.

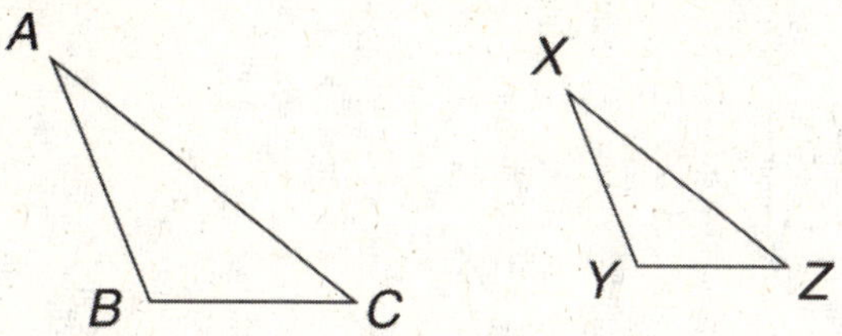

1. *AB* corresponds to ______.

2. *BC* corresponds to ______.

3. *AC* corresponds to ______.

Identify the corresponding sides. Then use ratios to determine whether the triangles are similar.

4.

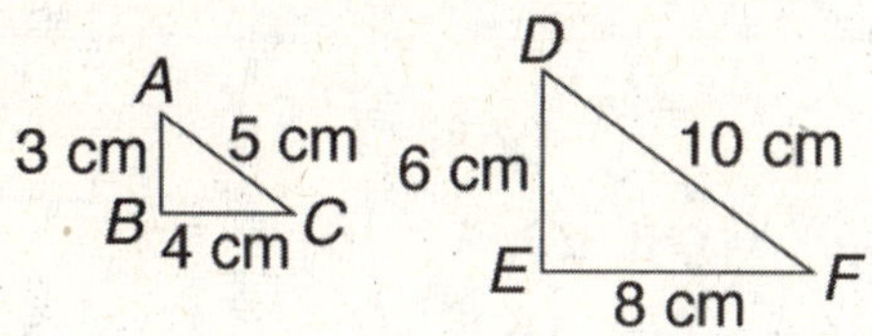

____

____

5.

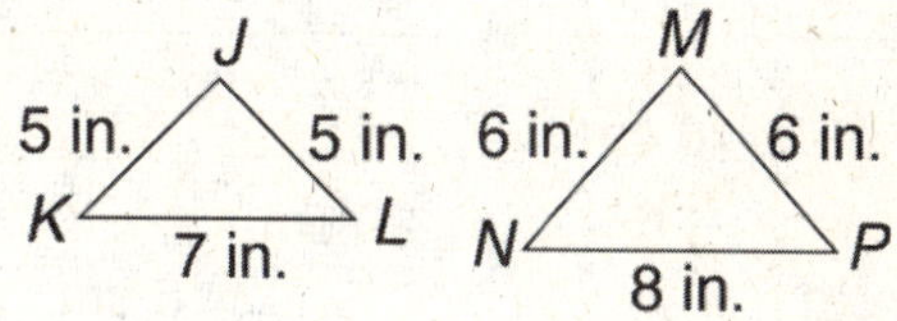

____

____

6.

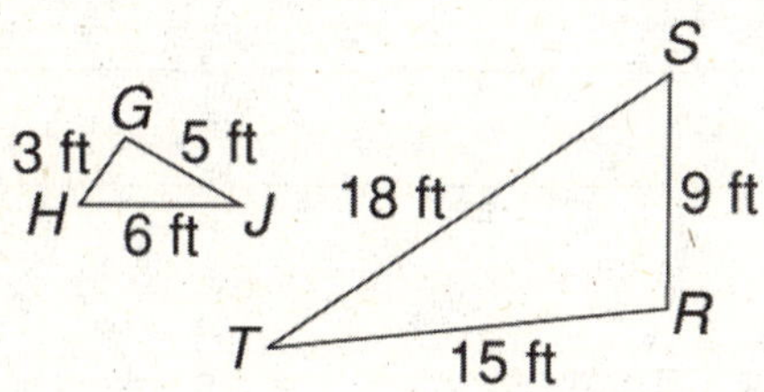

____

____

7.

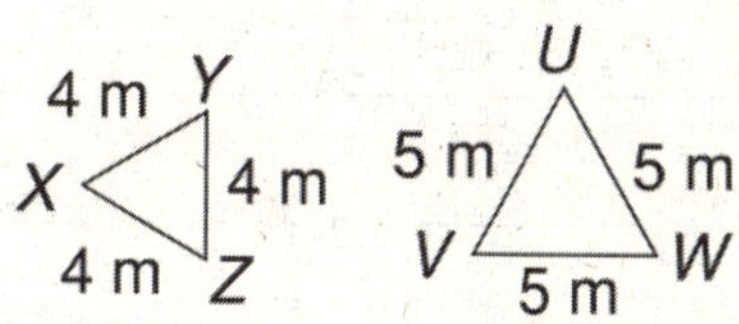

____

____

Use the properties of similarity to determine whether the figures are similar.

8.

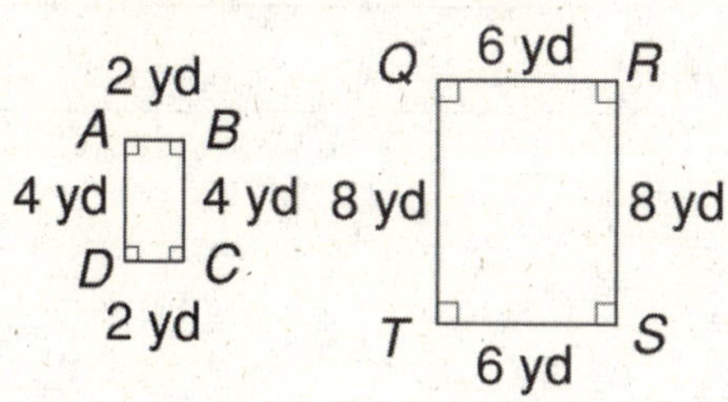

____

____

9

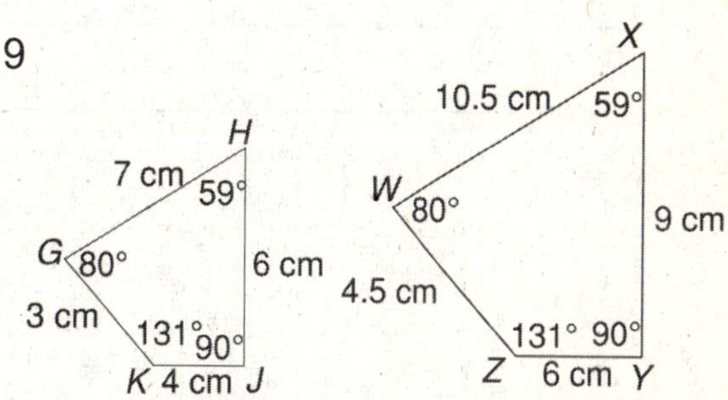

____

____

Name ______________________ Date ______________ Class ______________

LESSON 4-8

Practice B

Similar Figures and Proportions

Tell whether the triangles are similar.

1.

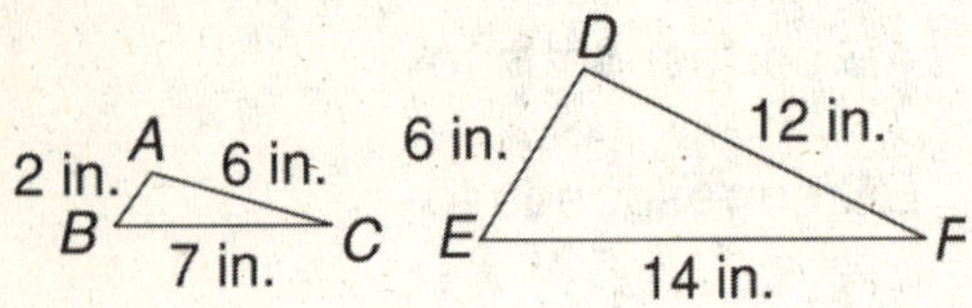

____

____

2.

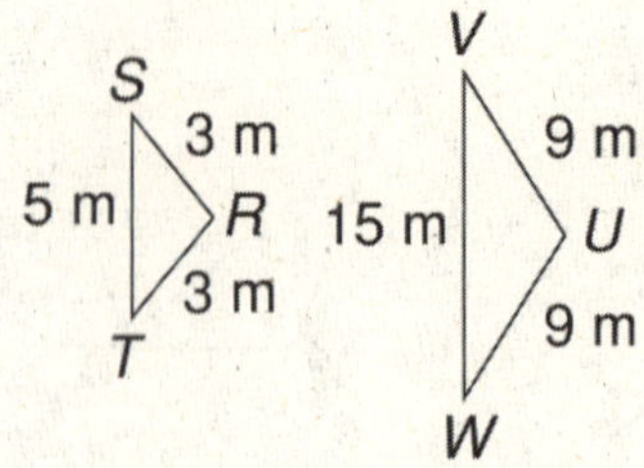

____

____

3.

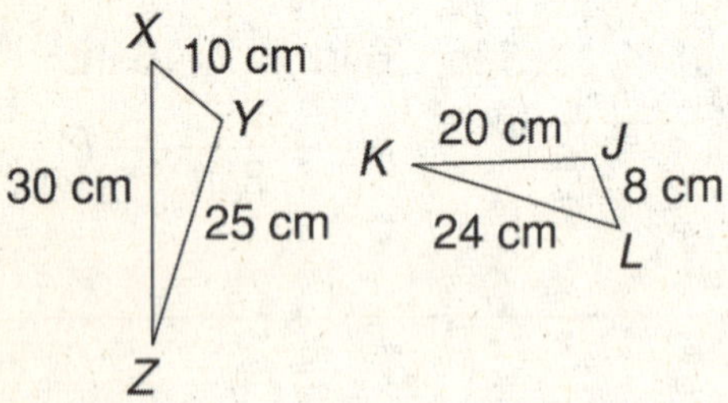

____

____

4.

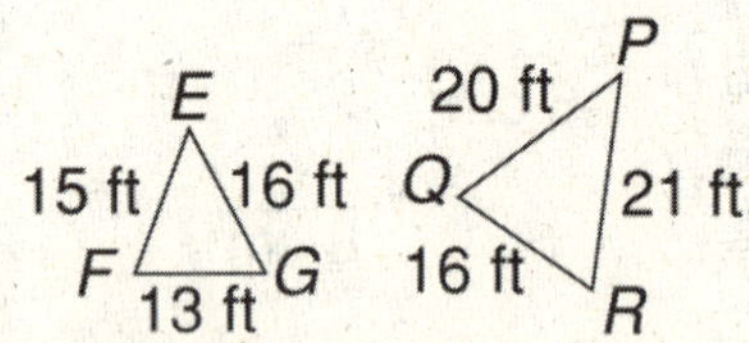

____

____

Tell whether the figures are similar.

5.

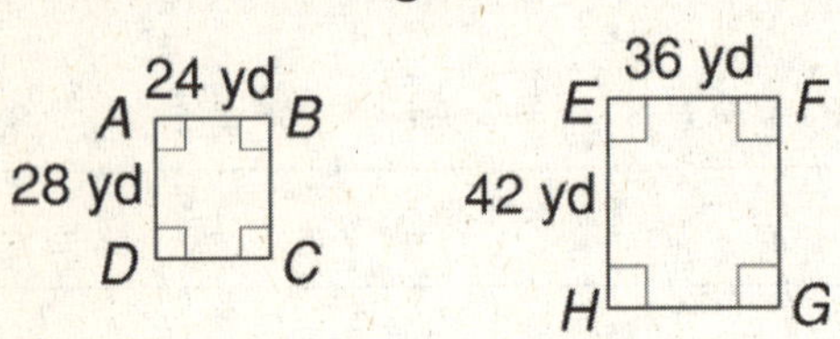

____

____

6.

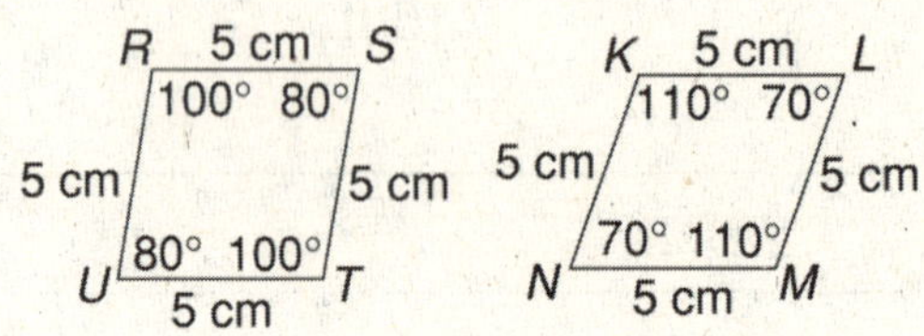

____

____

Name ______________________ Date ______________ Class ______________

LESSON 4-8

Practice C

Similar Figures and Proportions

Tell whether the figures are similar.

1.

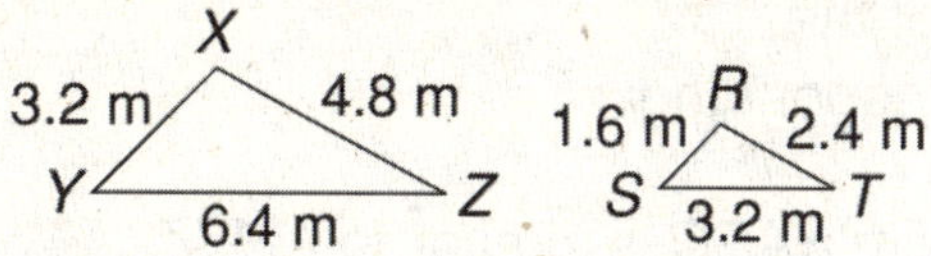

____

____

2.

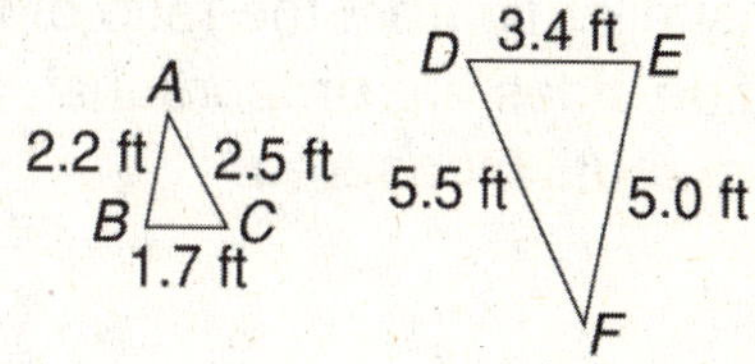

____

____

3.

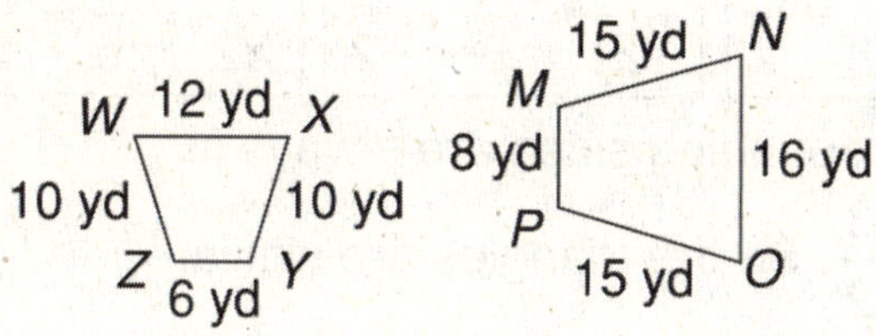

____

____

4.

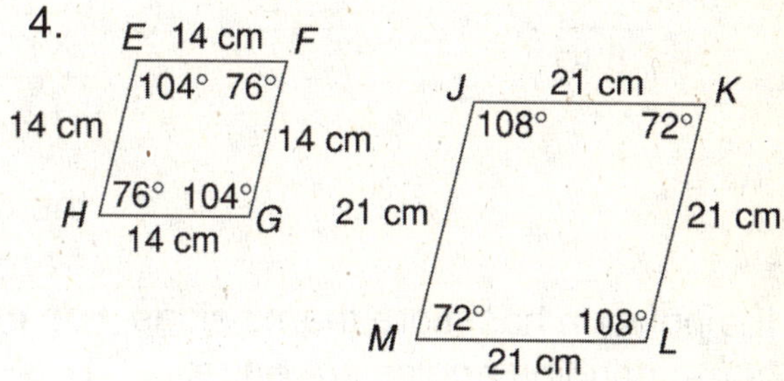

____

____

The figures in each pair are similar. Find the missing lengths or angle measures.

5.

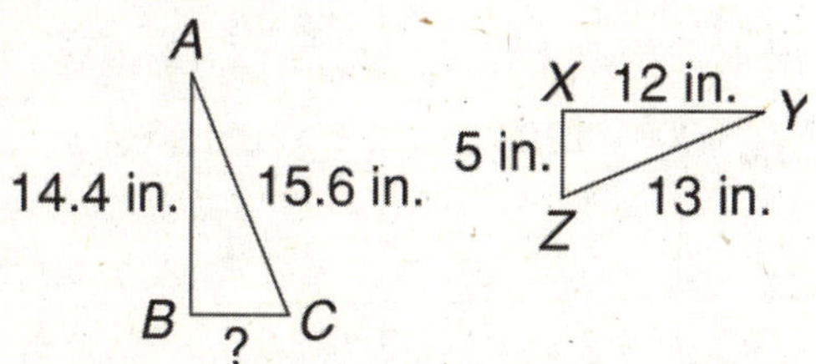

____

6.

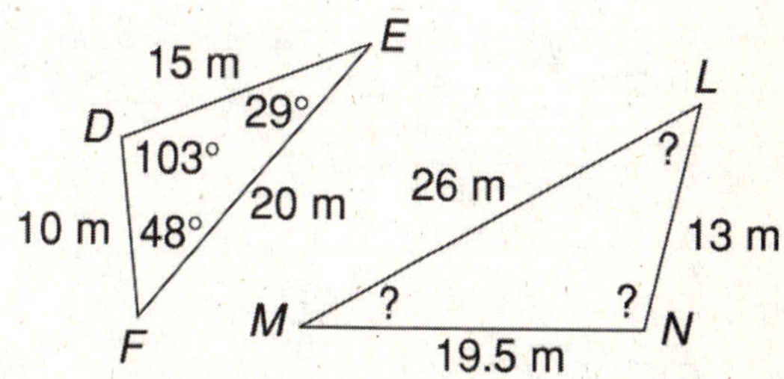

____

7.

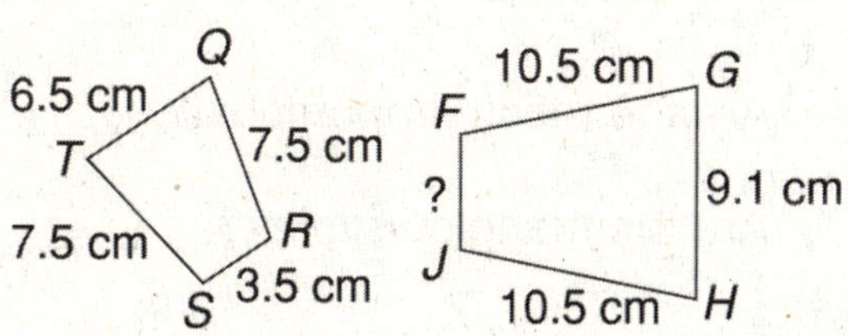

____

____

8.

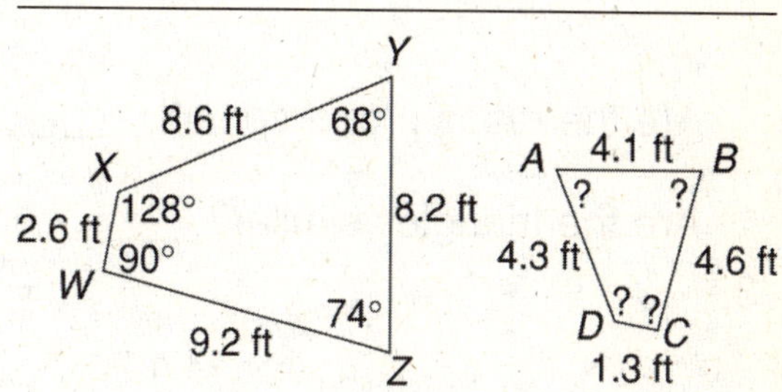

____

____

Name ______________________ Date ______________ Class ______________

LESSON 4-8

Review for Mastery

Similar Figures and Proportions

Figures that have the same shape but not the same size are **similar figures**. In similar figures, the ratio of the lengths of the corresponding sides are proportional, and the corresponding angles have equal measures.

To determine if ΔABC is similar to ΔXYZ, you can write a proportion for each pair of corresponding sides.

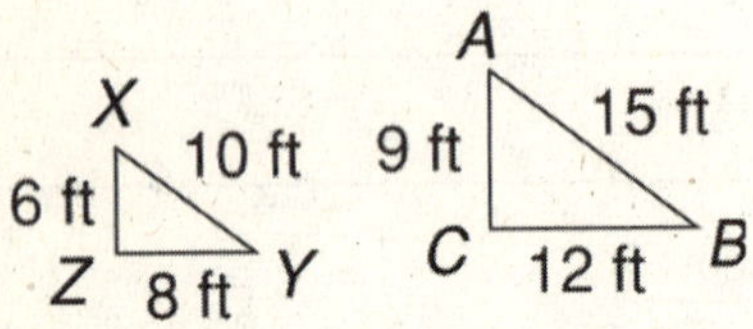

longest sides	middle sides	shortest sides
$\frac{AB}{XY} = \frac{15}{10} = \frac{3}{2}$	$\frac{BC}{YZ} = \frac{12}{8} = \frac{3}{2}$	$\frac{AC}{XZ} = \frac{9}{6} = \frac{3}{2}$

The corresponding sides are always in the ratio $\frac{3}{2}$. So the triangles are similar.

If a polygon has more than 3 sides, you must also show that the corresponding angles are equal.

Identify the corresponding sides. Use ratios to determine whether the figures are similar.

1. 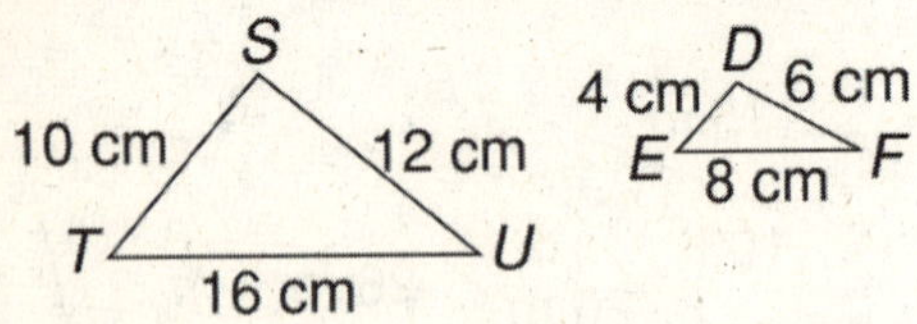

$\frac{TU}{EF} = \frac{\_\_}{8} = \frac{\_\_}{1}$; $\frac{SU}{\_\_} = \frac{\_\_}{\_\_} = \_\_$;

$\frac{ST}{\_\_} = \frac{\_\_}{\_\_} = \_\_$

Are the ratios proportional? ________

Are the triangles similar? ________

2. 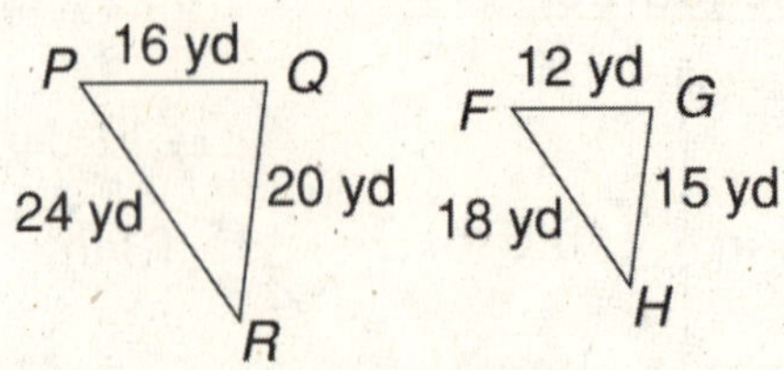

$\frac{PQ}{\_\_} = \frac{\_\_}{\_\_} = \_\_$; $\frac{PR}{\_\_} = \frac{\_\_}{\_\_} = \_\_$;

$\frac{QR}{\_\_} = \frac{\_\_}{\_\_} = \_\_$

Are the ratios proportional? ________

Are the triangles similar? ________

3. 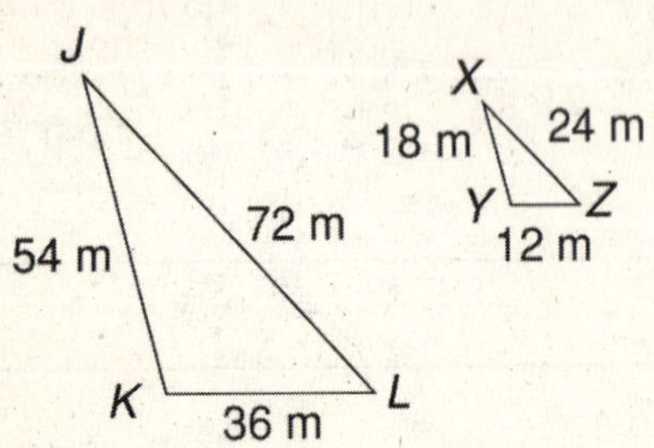

____

4. 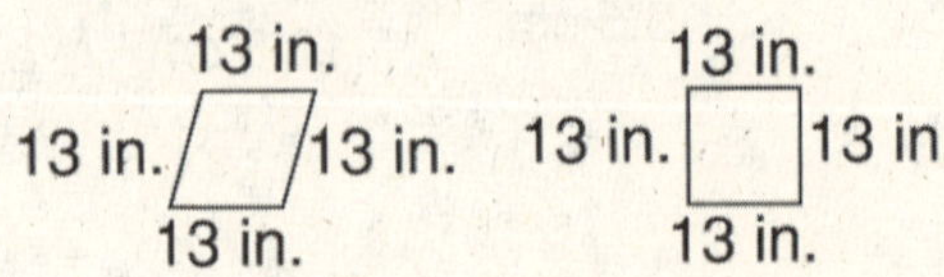

____

____

Name ____________________ Date ______________ Class ______________

LESSON 4-8

Challenge

The Same, Only Bigger

You can sometimes create a similar figure by using copies of the original figure.

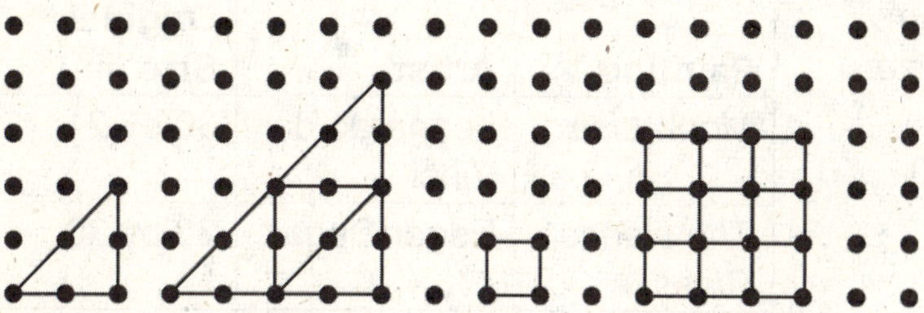

Original Figure — Similar Figure Scale factor = 2 — Original Figure — Similar Figure Scale factor = 3

Notice that the scale factor tells you how many times to repeat the original figure along each side or edge of the similar figure.

Use the given scale factor and copies of the original figure to draw a figure similar to the original figure.

1. 

scale factor = 4

2. 

scale factor = 2

3. 

scale factor = 3

4. 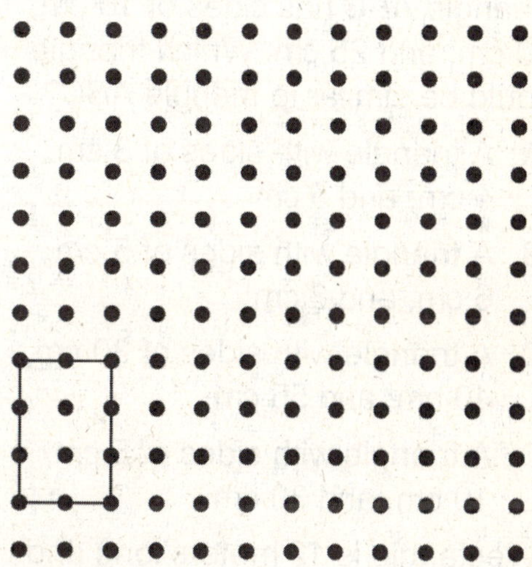

scale factor = 2

5. Draw a figure in the space below. Use a scale factor of 2 to create a similar figure.

Name ______________________ Date ______________ Class ______________

LESSON 4-8

Problem Solving

Similar Figures and Proportions

Use the information in the table to solve problems 1–3.

1. A small reproduction of one of the paintings in the list is similar in size. The reproduction measures 11 inches by 10 inches. Of which painting is this a reproduction?

____

Painting	**Artist**	**Original Size** (in.)
Mona Lisa	Leonardo da Vinci	30 by 21
The Dance Class	Edgar Degas	33 by 30
The Blue Vase	Paul Cézanne	22 by 18

2. A local artist painted a reproduction of Cézanne's painting. It measures 88 inches by 72 inches. Is the reproduction similar to the original? What is the ratio of corresponding sides?

____

3. A poster company made a poster of da Vinci's painting. The poster is 5 feet long and 3.5 feet wide. Is the poster similar to the original *Mona Lisa*? What is the ratio of corresponding sides?

____

Choose the letter for the best answer.

4. Triangle *ABC* has sides of 15 cm, 20 cm, and 25 cm. Which triangle could be similar to triangle *ABC*?

   A A triangle with sides of 3 cm, 4 cm, and 5 cm

   B A triangle with sides of 5 cm, 6 cm, and 8 cm

   C A triangle with sides of 30 cm, 40 cm, and 55 cm

   D A triangle with sides of 5 cm, 10 cm, and 30 cm

5. A rectangular picture frame is 14 inches long and 4 inches wide. Which dimensions could a similar picture frame have?

   F Length = 21 in.; width = 8 in.

   G Length = 35 in.; width = 15 in.

   H Length = 49 in.; width = 14 in.

   J Length = 7 in.; width = 3 in.

6. A rectangle is 12 meters long and 21 meters wide. Which dimensions correspond to a nonsimilar rectangle?

   A 4 m; 7 m   C 20 m; 35 m

   B 8 m; 14 m   D 24 m; 35 m

7. A rectangle is 6 feet long and 15 feet wide. Which dimensions correspond to a similar rectangle?

   F 8 ft; 24 ft   H 15 ft; 35 ft

   G 10 ft; 25 ft   J 18 ft; 40 ft

Name ______________________ Date ______________ Class ______________

LESSON 4-8

Reading Strategies

Understanding Vocabulary

Similar means almost the same. If two objects are similar, they have some things in common.

Similar figures are figures that are nearly the same. Similar figures have the same shape, but are different sizes.

Similar figures have **corresponding sides** and **corresponding angles**. *Corresponding* means matching. Each side and angle in a similar figure has a corresponding side and angle.

These two triangles are similar.

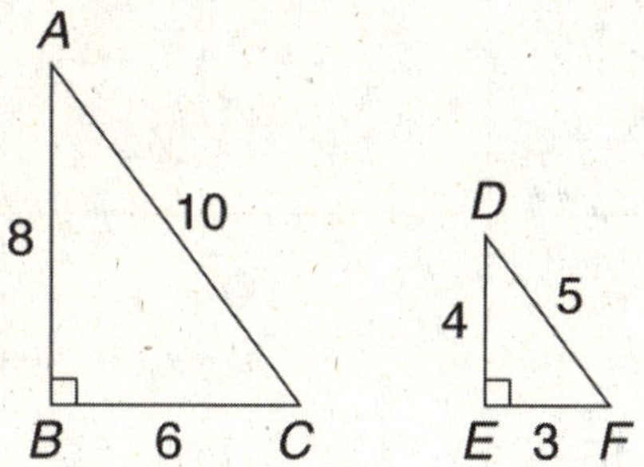

Use the figures above to answer each question.

1. What angle corresponds to angle *B*?

____

2. What angle corresponds to angle *A*?

____

3. What side corresponds to side *BC*?

____

Are the figures similar? Answer yes or no for each pair.

4. ____________

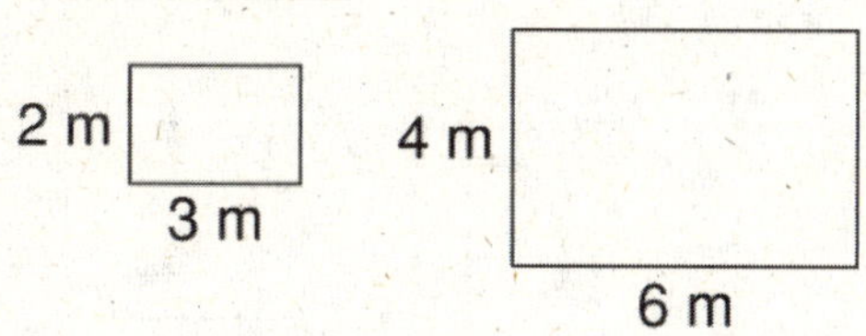

5. ____________

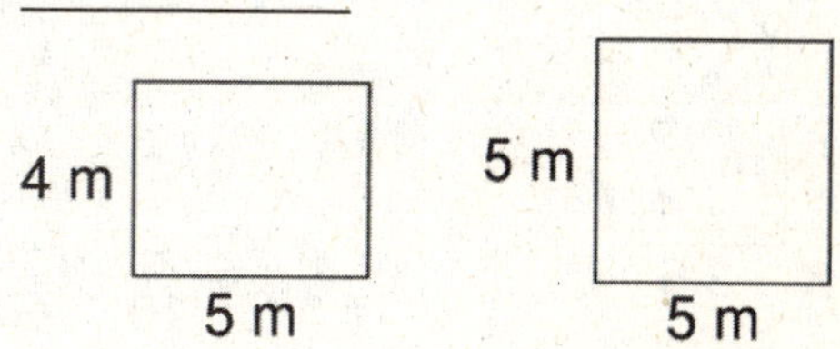

Name ______________________ Date ______________ Class ______________

LESSON 4-8

Puzzles, Twisters & Teasers

Concentrating on Figures

Pretend this is a game of concentration. The object of the game is to match cards with similar figures. Each box represents a card with a figure on it. When you match 2 cards, cross them out. Rearrange the letters of the unmatched cards to solve the riddle.

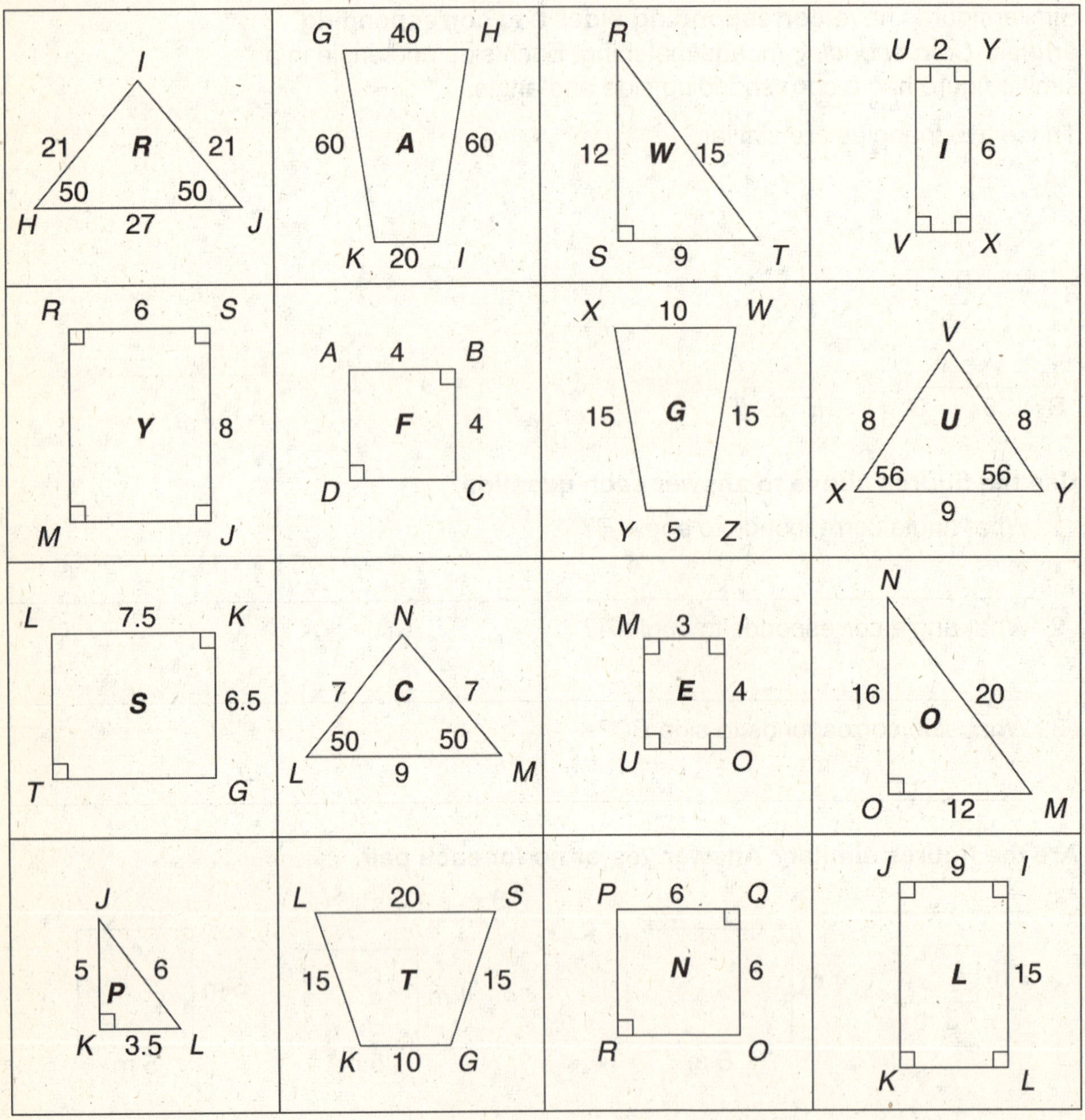

What kind of flowers are on your face?

___ ___ ___ ___ ___ ___

Name ______________________ Date ______________ Class ______________

LESSON 4-9

Practice A

Using Similar Figures

For each pair of similar figures write a proportion containing the unknown length. Then solve.

1.

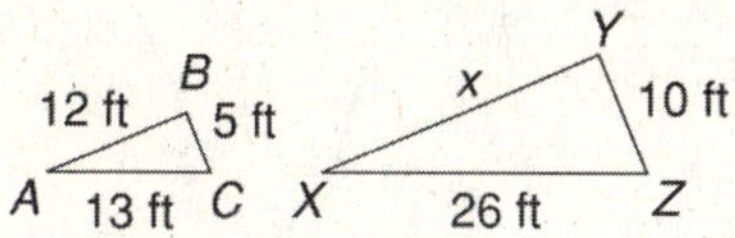

____

2.

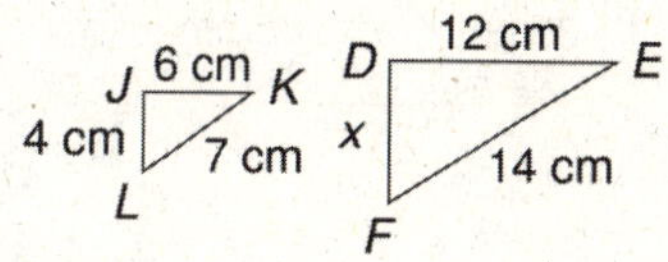

____

3.

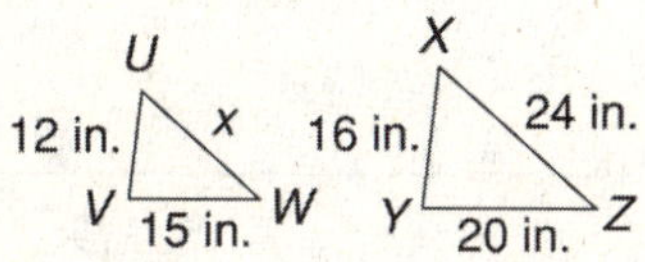

____

4.

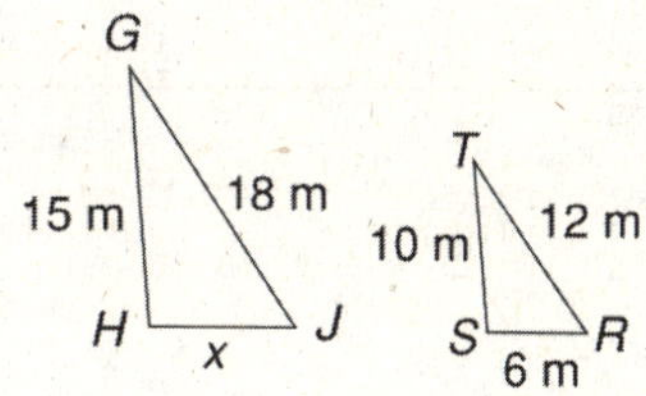

____

5. Kareem and Julio have rectangular model train layouts that are similar to each other. Julio's layout is 4 feet by 7 feet. Kareem's layout is 6 feet wide. What is the length of Kareem's layout?

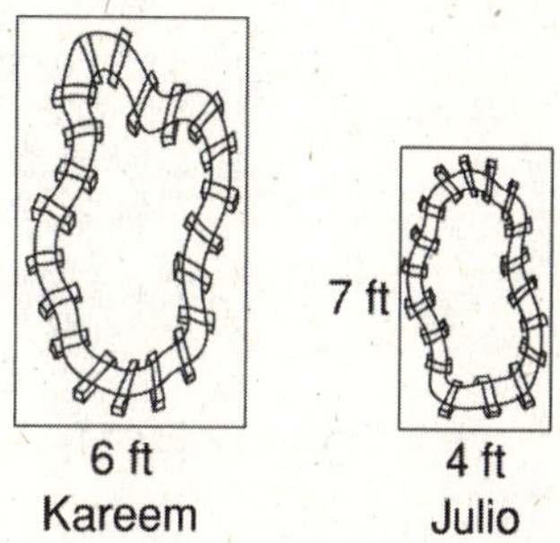

____

6. A 6-foot-tall adult casts a shadow that is 15 feet long. Estimate the height of a child who casts a 10-foot shadow.

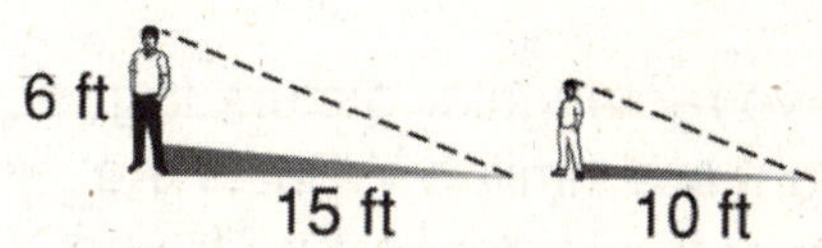

____

Name ______________________ Date ______________ Class ______________

LESSON 4-9

Practice B

Using Similar Figures

$\triangle ABC \sim \triangle DEF$ in each pair. Find the unknown measures.

1.

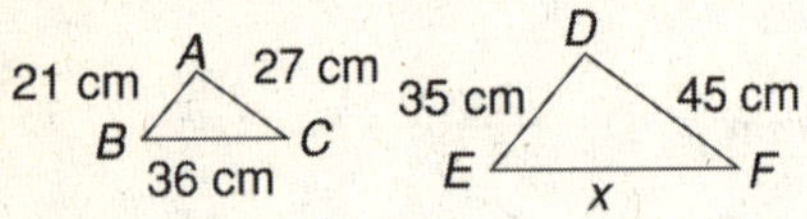

____

2.

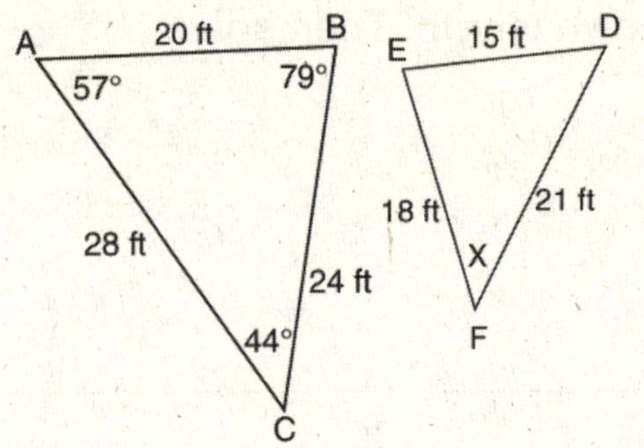

____

3.

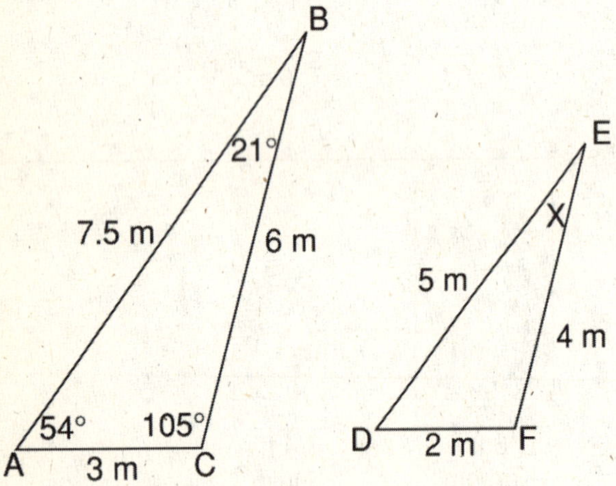

4.

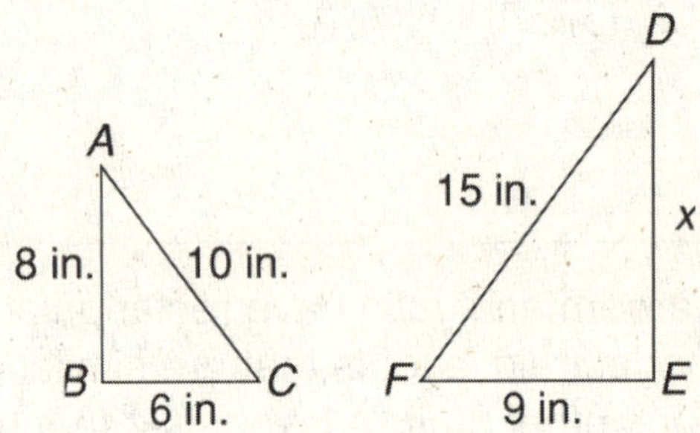

____

5. The two rectangular picture frames at the right are similar. What is the height of the larger picture frame?

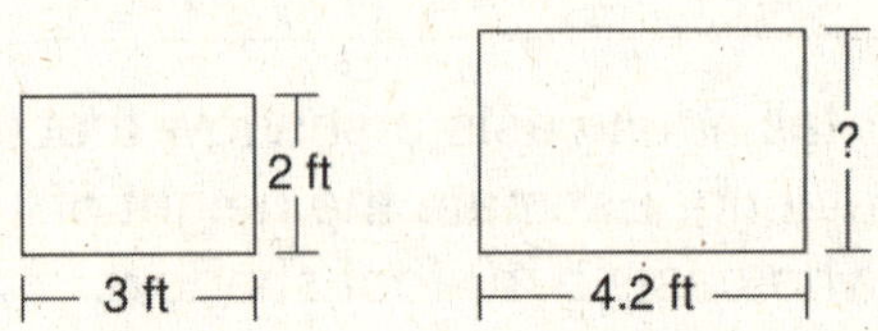

____

6. A palm tree casts a shadow that is 44 feet long. A 6-foot ladder casts a shadow that is 16 feet long. Estimate the height of the palm tree.

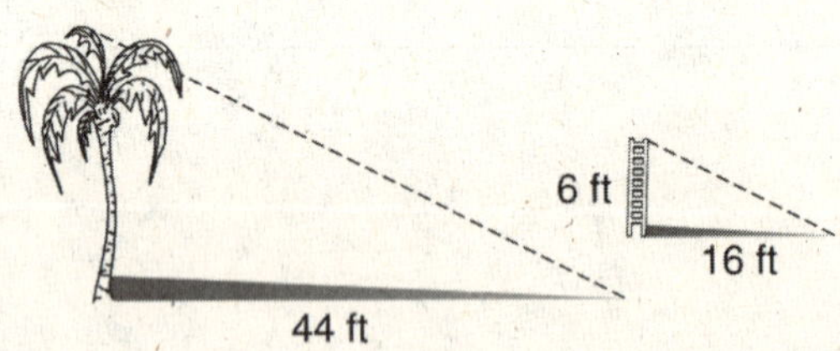

____

Name ______________________ Date ______________ Class ______________

LESSON 4-9

Practice C

Using Similar Figures

Find the unknown measure in each pair of similar figures.

1.

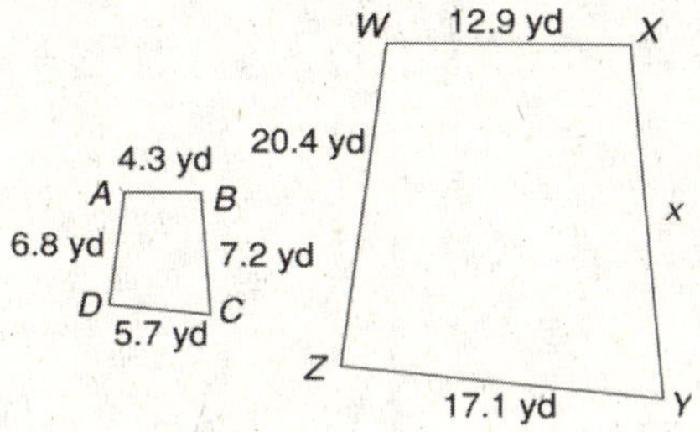

____

2.

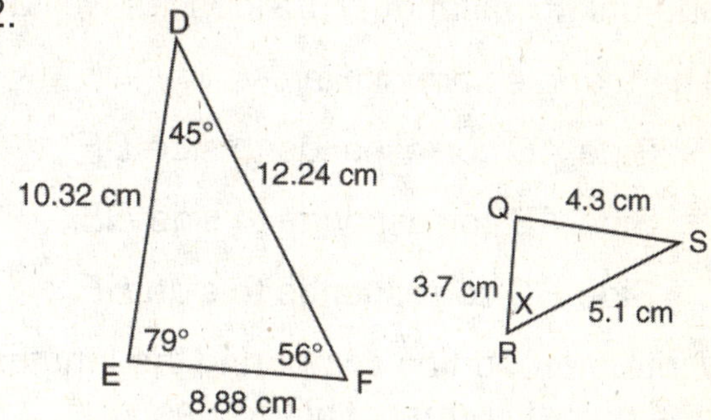

____

3.

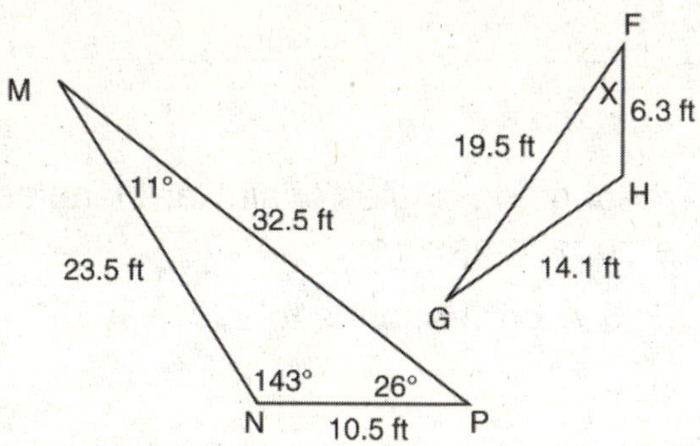

____

4.

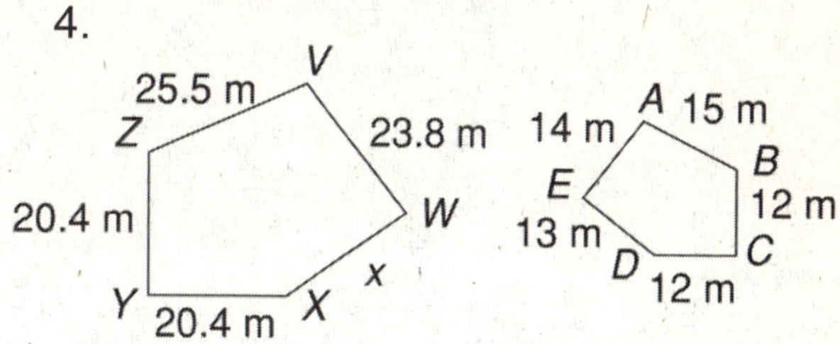

____

Estimate the height of each object in the picture below.

5. house ______________

6. tree ______________

7. lamppost ______________

8. radio tower ______________

Name ______________________ Date ______________ Class ______________

LESSON 4-9

Review for Mastery

Using Similar Figures

If you know that 2 figures are similar, you can use proportions to find unknown lengths of sides, as well as find unknown angle measurements.

The triangles are similar.

Side *AC* corresponds to side *DF*.

Side *AB* corresponds to side *DE*.

Side *BC* corresponds to side *EF*.

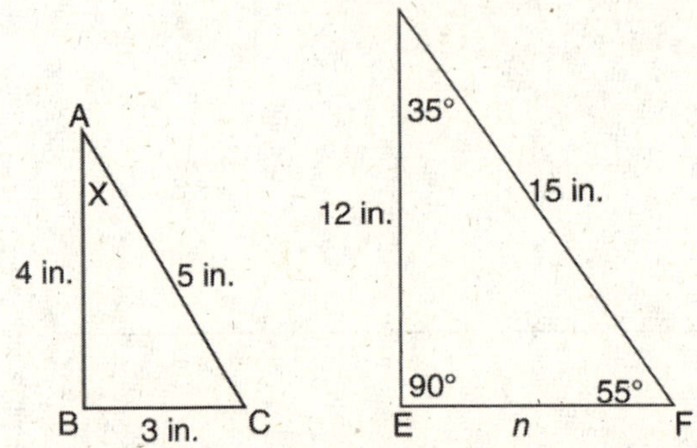

Write a proportion comparing the lengths of a pair of corresponding sides.

$$\frac{AC}{DF} = \frac{BC}{EF}$$

$$\frac{5}{15} = \frac{3}{n}$$

$$5 \cdot n = 15 \cdot 3$$

$$5n = 45$$

$$\frac{5n}{5} = \frac{45}{5}$$

$$n = 9$$

The length of the missing side is 9 in.

Corresponding angles of similar triangles have equal measures.

Since $\angle A$ corresponds to $\angle D$, $x = 35°$.

The measure of the unknown angle is 35°.

Find the unknown measure in each pair of similar figures.

1.

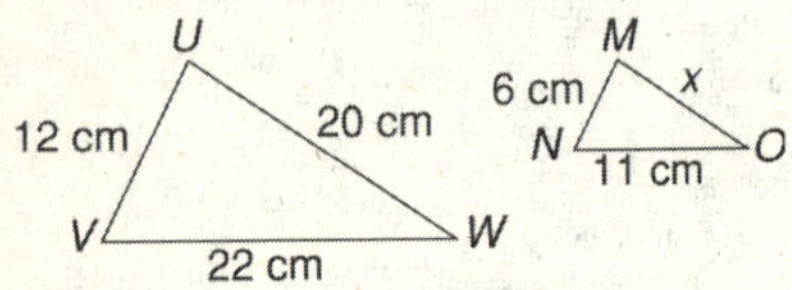

$\frac{UW}{\_\_} = \frac{UV}{\_\_}$; $\frac{20}{\_\_} = \frac{12}{\_\_}$

x = ______________________

2.

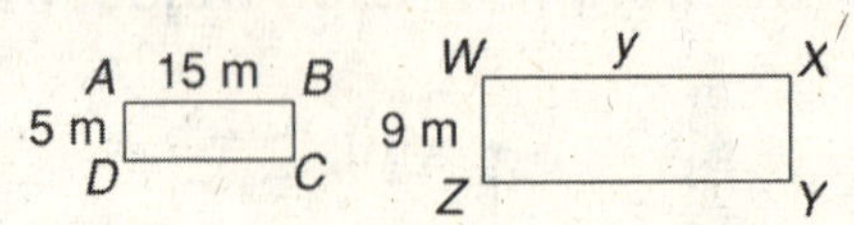

$\frac{WZ}{\_\_} = \frac{WX}{\_\_}$; $\frac{9}{\_\_} = \_\_$

y = ______________________

3.

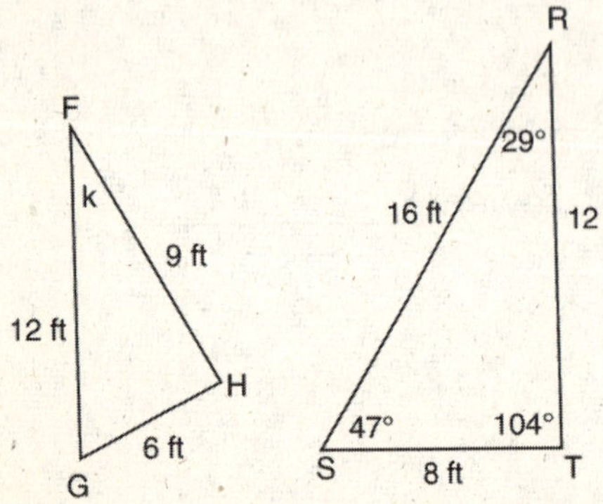

k = ______________________

4.

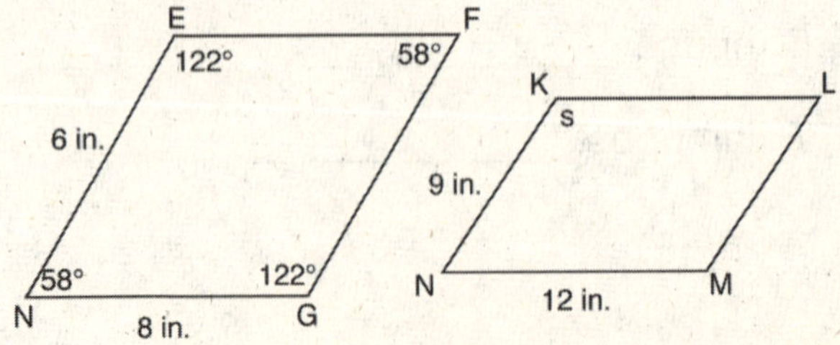

s = ______________________

Name ______________________ Date ______________ Class ______________

LESSON 4-9

Challenge

You Be the Artist

Your club is planning a poster to advertise the school's international dinner. The poster will be enlarged and used as a mural on the school cafeteria wall. The poster will also be reduced and used as flyers. The mural will be 10 feet high and 15 feet long. The flyers will be printed on $8\frac{1}{2}$-by-11-inch paper.

Plan the size of the poster so that the enlargement and reduction will be easy to make.

1. What is the width-to-length ratio for the wall mural? Write the ratio in simplest terms.

____

2. What is the width-to-length ratio for the flyer? Write the ratio in simplest terms.

____

3. Do you want your artwork to fill the entire page for the flyer?

____

4. What are some possible dimensions for your poster?

____

5. Will your poster fill the wall space when it is enlarged for the mural? Explain.

____

6. What would be a good size for the artwork on the flyer?

____

Name ______________________ Date ______________ Class ______________

LESSON 4-9

Problem Solving

Using Similar Figures

Write the correct answer.

1. An architect is building a model of a tennis court for a new client. On the model, the court is 6 inches wide and 13 inches long. An official tennis court is 36 feet wide. What is the length of a tennis court?

____

2. Mr. Hemley stands next to the Illinois Centennial Monument at Logan Square in Chicago and casts a shadow that is 18 feet long. The shadow of the monument is 204 feet long. If Mr. Hemley is 6 feet tall, how tall is the monument?

____

3. The official size of a basketball court in the NBA is 94 feet by 50 feet. The basketball court in the school gym is 47 feet long. How wide must it be to be similar to an NBA court?

____

4. Two rectangular desks are similar. The larger one is 42 inches long and 18 inches wide. The smaller one is 35 inches long. What is the width of the smaller desk?

____

Choose the letter for the best answer.

5. An isosceles triangle has two sides that are equal in length. Isosceles triangle *ABC* is similar to isosceles triangle *XYZ*. What proportion would you use to find the length of the third side of triangle *XYZ*?

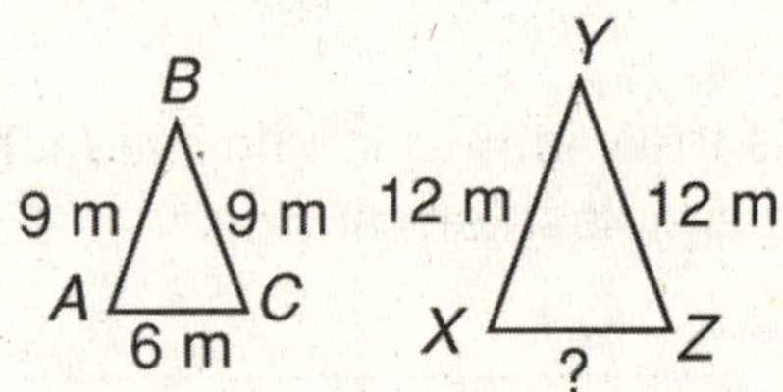

A $\frac{BC}{XZ} = \frac{AB}{XY}$

B $\frac{AC}{XY} = \frac{BC}{XZ}$

C $\frac{AB}{XY} = \frac{AC}{XZ}$

D $\frac{AB}{XY} = \frac{BC}{YZ}$

6. The dining room at Monticello, Thomas Jefferson's home in Virginia, is 216 inches by 222 inches. Of the following, which size rug would be similar in shape to the dining room?

F 72 inches by 74 inches

G 108 inches by 110 inches

H 118 inches by 111 inches

J 84 inches by 96 inches

7. A 9-foot street sign casts a 12-foot shadow. The lamppost next to it casts a 24-foot shadow. How tall is the lamppost?

A 24 feet

B 15 feet

C 18 feet

D 36 feet

Name ______________________ Date ____________ Class ____________

LESSON 4-9

Reading Strategies

Use a Flowchart

It is very difficult to measure the height of tall tree or a utility pole directly. You can set up proportions to measure very tall objects indirectly.

This method of measuring is called **indirect measurement**. You do not actually measure the height. You use a proportion to find the height.

These two triangles are similar.
Find length x in triangle DEF.

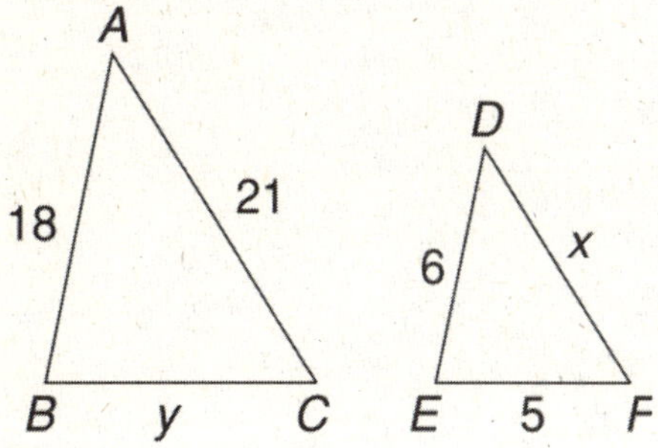

The flowchart helps you set up a proportion to find the value of x.

Write a proportion to find the missing length.

$$\frac{AB}{DE} = \frac{AC}{DF}$$

↓

Put the lengths of the known sides into the equation. x stands for the length of the unknown side.

$$\frac{18}{6} = \frac{21}{x}$$

↓

Multiply cross products.
$18 \cdot x = 6 \cdot 21$

↓

Find the value of x.

$$\frac{18x}{18} = \frac{6 \cdot 21}{18}$$

$$x = \frac{126}{18}$$

$$x = 7$$

Answer each question.

1. Why is this method called indirect measurement?

____

____

2. What is the next step after setting up the proportion?

____

3. Write a proportion to find the length y in triangle ABC.

____

Name ______________________ Date ______________ Class ______________

LESSON 4-9

Puzzles, Twisters & Teasers

Measure This!

Below are two similar rectangles. Use indirect measurement to find the width of the smaller rectangle. Show the steps you used by filling in the blanks in the sentences. To solve the riddle, find the letter(s) in each answer with a number below it. Match the letters to the numbered blanks in the riddle.

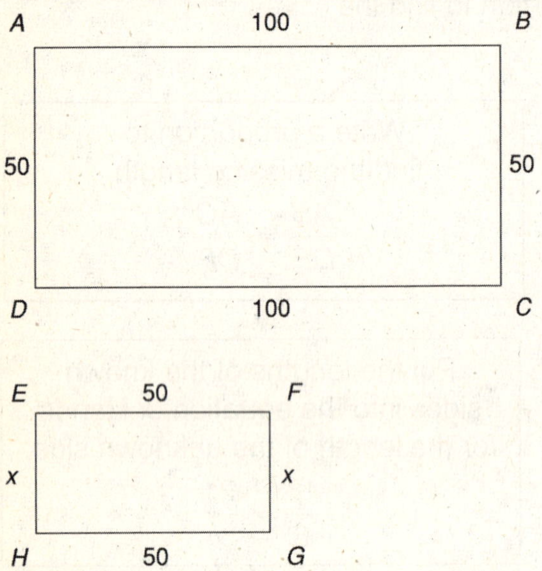

1. Write a ___ ___ ___ ___ ___ ___ ___ ___ ___ ___.
(blank 3 numbered 2)

2. Substitute the ___ ___ ___ ___ ___ ___ of the sides.
(blank 5 numbered 6)

3. Find the ___ ___ ___ ___ ___ ___ ___ ___ ___ ___ ___ ___ ___.
(blank 1 numbered 4; blank 6 numbered 5; blank 10 numbered 3)

4. ___ ___ ___ ___ ___ for x.
(blank 1 numbered 1)

Divide.

5. $x =$ ____

What do you call a cat that swims and has eight legs?

An ___ ___ ___ ___ ___ ___ ___ ___
2 4 6 2 5 3 1 1

Name ______________________ Date ______________ Class ______________

LESSON 4-10

Practice A

Scale Drawings and Scale Models

Identify the scale factor. Choose the best answer.

1. Person: 72 inches
   Action figure: 6 inches

   A $\frac{1}{7}$ C $\frac{1}{12}$

   B $\frac{1}{10}$ D $\frac{1}{15}$

2. Dog: 24 inches
   Stuffed animal: 8 inches

   F $\frac{1}{3}$ H $\frac{1}{5}$

   G $\frac{1}{4}$ J $\frac{1}{6}$

3. Fish: 16 inches
   Fishing lure: 2 inches

   A $\frac{1}{6}$ C $\frac{1}{12}$

   B $\frac{1}{8}$ D $\frac{1}{14}$

4. House: 30 feet
   Dollhouse: 3 feet

   F $\frac{1}{3}$ H $\frac{1}{27}$

   G $\frac{1}{10}$ J $\frac{1}{33}$

Identify the scale factor.

5.

	Guitar	Ukulele
Length (in.)	36	18

____

6.

	Car	Toy Car
Length (ft)	12	3

____

7.

	Flute	Piccolo
Length (in.)	30	10

____

8.

	Poodle	Toy Poodle
Height (in.)	56	8

____

9. On a road map of New York, the distance from New York City to Albany is 3 inches. The map scale is 1 in:50 mi. What is the actual distance between the cities?

____

10. On a scale drawing, a bookshelf is 8 inches tall. The scale factor is $\frac{1}{8}$. What is the height of the bookshelf?

____

Name ______________________ Date ______________ Class ______________

LESSON 4-10

Practice B

Scale Drawings and Scale Models

Identify the scale factor.

1.

	Alligator	**Toy Alligator**
Length (in.)	175	7

____

2.

	Airplane	**Model**
Length (ft)	24	3

____

3.

	Car	**Toy Car**
Length (ft.)	13.5	1.5

____

4.

	Person	**Action Figure**
Height (in.)	66	6

____

5.

	Boat	**Model**
Length (in.)	128	8

____

6.

	Fish	**Fishing Lure**
Length (in.)	18	2

____

7.

	Tiger	**Stuffed Animal**
Length (in.)	70	14

____

8.

	House	**Dollhouse**
Height (ft)	39.2	2.8

____

9. On a scale drawing, a school is 1.6 feet tall. The scale factor is $\frac{1}{22}$. Find the height of the school. ______________

10. On a road map of Pennsylvania, the distance from Philadelphia to Washington, D.C., is 6.8 centimeters. The map scale is 2 cm:40 mi. What is the actual distance between the cities? ______________

11. On a scale drawing, a bicycle is $6\frac{4}{5}$ inches tall. The scale factor is $\frac{1}{6}$. Find the height of the bicycle. ______________

Name ______________________ Date ______________ Class ______________

LESSON 4-10

Practice C

Scale Drawings and Scale Models

Identify the scale factor.

1.

	Bear	**Stuffed Animal**
Height (in.)	62	15.5

____

2.

	House	**Dollhouse**
Height (ft)	32.4	2.7

____

3.

	Airplane	**Model**
Length (ft)	25.5	1.5

____

4.

	Alligator	**Toy Alligator**
Length (in.)	128.1	6.1

____

The scale factor of each model is 1:16. Find the missing dimensions.

Item	**Actual Dimensions**	**Model Dimensions**
5. barn	height: 32 ft length:	height: length: 3.5 ft
6. submarine	length:	length: $18\frac{3}{4}$ ft
7. bookcase	height: 96 in.	height:
8. tree	height:	height: $2\frac{1}{2}$ ft
9. car	length: 13 ft height: 5.5 ft	length: height:
10. shark	length:	length: $14\frac{1}{4}$ in.

11. Hillary took a photograph of her house, which has an actual height of 28.5 feet. If the house measures 3.6 inches tall in the photograph, what is the scale factor? ______________

12. On a road map, the distance from Portland to Seattle is 8 centimeters. The map scale is 2 cm:37.5 mi. What is the actual distance between the cities? ______________

13. A sculptor plans a statue by making a drawing to scale. On the drawing, the statue is $8\frac{2}{5}$ inches tall. The scale factor in the drawing is $\frac{1}{23}$. Find the height of the statue. ______________

Name ______________________ Date ______________ Class ______________

LESSON 4-10

Review for Mastery

Scale Drawings and Scale Models

The dimensions of a scale model or scale drawing are related to the actual dimensions by a *scale factor*. The **scale factor** is a ratio.

The length of a model car is 9 in. →
The length of the actual car is 162 in. →

$$\frac{9\text{ in.}}{162\text{ in.}} = \frac{9 \div 9}{162 \div 9} = \frac{1}{18}$$

$\frac{9}{162}$ can be simplified to $\frac{1}{18}$. — The scale factor is $\frac{1}{18}$.

If you know the scale factor, you can use a proportion to find the dimensions of an actual object or of a scale model or drawing.

- The scale factor of a model train set is $\frac{1}{87}$. A piece of track in the model train set is 8 in. long. What is the actual length of the track?

  $\frac{\text{model length}}{\text{actual length}} = \frac{8}{x}$ $\quad \frac{8}{x} = \frac{1}{87}$ $\quad x = 696$

  The actual length of track is 696 inches.

- The distance between 2 cities on a map is 4.5 centimeters. The map scale is 1 cm: 40 mi.

  $\frac{\text{distance on map}}{\text{actual distance}} = \frac{4.5\text{ cm}}{x\text{ mi}} = \frac{1\text{ cm}}{40\text{ mi}}$ $\quad \frac{4.5}{x} = \frac{1}{40}$ $\quad x = 180$

  The actual distance is 180 miles.

Identify the scale factor.

1. Photograph: height 3 in.
   Painting: height 24 in.

   $\frac{\text{photo height}}{\text{painting height}} = \frac{\_\_\_\text{ in.}}{\_\_\_\text{ in.}} = \frac{\_\_\_}{\_\_\_}$

2. Butterfly: wingspan 20 cm
   Silk butterfly: wingspan 4 cm

   $\frac{\text{silk butterfly}}{\text{butterfly}} = \frac{\_\_\_\text{ cm}}{\_\_\_\text{ cm}} = \frac{\_\_\_}{\_\_\_}$

3. On a scale drawing, the scale factor is $\frac{1}{12}$. A plum tree is 7 inches tall on the scale drawing. What is the actual height of the tree?

   ____

4. On a road map, the distance between 2 cities is 2.5 inches. The map scale is 1 inch:30 miles. What is the actual distance between the cities?

   ____

Name ______________________ Date ______________ Class ______________

LESSON 4-10

Challenge

Balls of Sports

Each circle below is a scale drawing of a different type of ball used in a sport.

- Measure the diameter of each circle to the nearest tenth of a centimeter.
- Use the scale to find the actual diameter of the ball to the nearest tenth of a centimeter.
- Use the chart below to find the sport in which the ball is used.

Diameter of Balls Used in Various Sports	
Basketball	24.0 cm
Baseball	7.5 cm
Golf	4.2 cm
Table Tennis	3.8 cm
Tennis	6.4 cm
Volleyball	21.0 cm

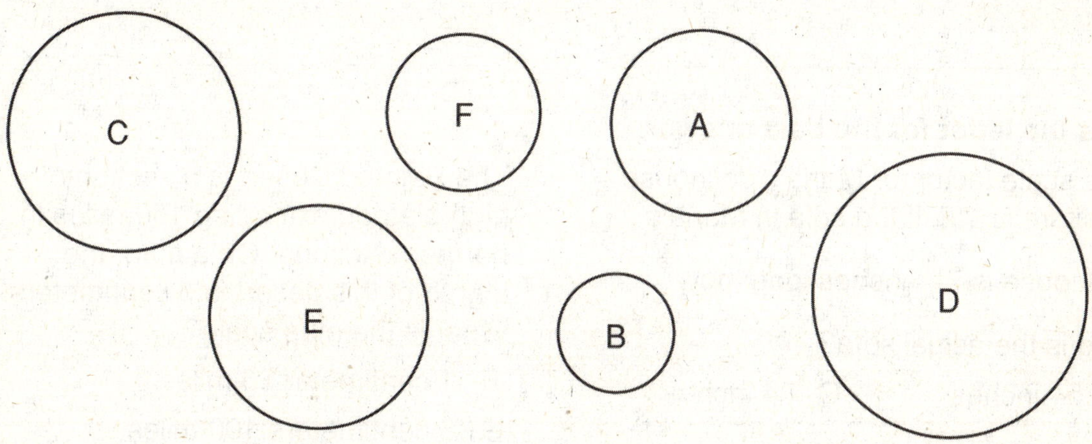

	Circle	Scale	Measured Diameter	Actual Diameter	Sport
1.	A	1 cm = 3 cm			
2.	B	1 cm = 15 cm			
3.	C	1 cm = 2 cm			
4.	D	1 cm = 1 cm			
5.	E	1 cm = 1.4 cm			
6.	F	1 cm = 10 cm			

Name ______________________ Date ______________ Class ______________

LESSON 4-10

Problem Solving

Scale Drawings and Scale Models

Write the correct answer.

1. The scale on a road map is 1 cm:500 mi. If the distance on the map between New York City and Memphis is 2.2 centimeters, what is the actual distance between the two cities?

____

2. There are several different scales in model railroading. Trains designated as O gauge are built to a scale factor of 1:48. To the nearest hundredth of a foot, how long is a model of a 50-foot boxcar in O gauge?

____

3. For a school project, LeeAnn is making a model of the Empire State Building. She is using a scale of 1 cm:8 ft. The Empire State Building is 1,252 feet tall. How tall is her model?

____

4. A model of the Eiffel Tower that was purchased in a gift shop is 29.55 inches tall. The actual height of the Eiffel Tower is 985 feet, or 11,820 inches. What scale factor was used to make the model?

____

Choose the letter for the best answer.

5. The scale factor for Maria's dollhouse furniture is 1:8. If the sofa in Maria's dollhouse is $7\frac{1}{2}$ inches long, how long is the actual sofa?

A 54 inches
B 60 inches
C 84 inches
D $15\frac{1}{2}$ inches

6. The Painted Desert is a section of high plateau extending 150 miles in northern Arizona. On a map, the length of this desert is 5 centimeters. What is the map scale?

F 1 centimeter:25 miles
G 5 centimeters:100 miles
H 1 centimeter:30 miles
J 1 centimeter:50 miles

7. Josh wants to add a model of a tree to his model railroad layout. How big should the model tree be if the actual tree is 315 inches and the scale factor is 1:90?

A 395 inches
B 39.5 inches
C 35 inches
D 3.5 inches

8. The scale on a wall map is 1 in:55 mi. What is the distance on the map between two cities that are 99 miles apart?

F 44 inches
G 1.8 inches
H 2.5 inches
J 0.55 inches

Name ______________________ Date ______________ Class ______________

LESSON 4-10

Reading Strategies

Read a Map

A **scale drawing** has the same shape, but is not the same size, as the object it represents. A map is an example of a scale drawing.

This is a map of a campground. The scale is 1 cm:10 ft.

To find how far the campground entrance is from the canoe rental office, follow the steps. Use a centimeter ruler to measure.

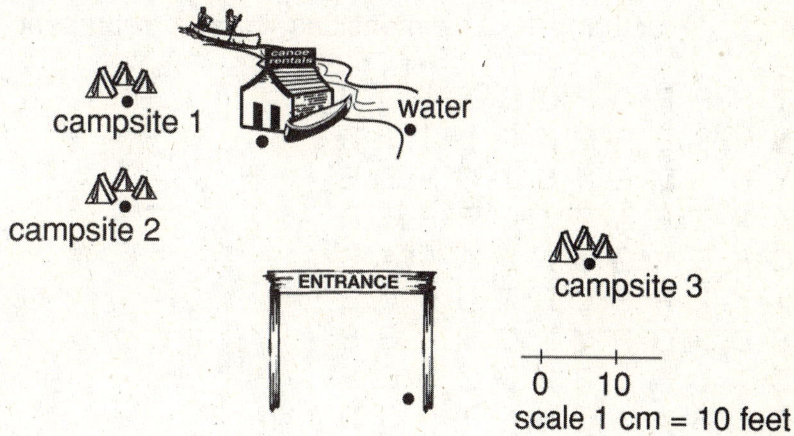

Step 1: Measure the distance in centimeters.

→ The distance is 4 centimeters.

Step 2: Set up a proportion using the map scale as one ratio.

→ $\frac{1 \text{ cm}}{10 \text{ ft}} = \frac{4 \text{ cm}}{x \text{ ft}}$

Step 3: Set up cross products. → $1x = 4 \cdot 10$

Step 4: Solve to find the value of x. → $x = 40$

Use the map to answer each question.

1. How many centimeters is Campsite 3 from the water?

____

2. Write a proportion to find the distance from Campsite 3 to the water.

____

3. How many centimeters is Campsite 3 from the canoe rental office?

____

4. Write a proportion to find the distance from Campsite 3 to the canoe rental office.

____

Name ______________________ Date ______________ Class ______________

LESSON 4-10

Puzzles, Twisters & Teasers

Let's Rock!

Find and circle these words in the word search. Find a word that solves the riddle. Circle it and write it on the line.

scale	model	drawing	factor	ratio
size	actual	dimensions	represent	proportion

V	T	S	P	R	O	P	O	R	T	I	O	N
M	N	V	S	B	R	M	K	O	L	J	I	R
O	Y	D	I	M	E	N	S	I	O	N	S	E
D	A	X	Z	R	A	T	I	O	F	F	C	P
E	O	P	E	V	O	Q	W	E	A	A	A	R
L	I	K	L	P	S	C	R	T	U	C	L	E
A	S	D	F	G	H	M	K	Q	R	T	E	S
A	C	T	U	A	L	B	V	E	Z	O	E	E
N	O	D	R	A	W	I	N	G	T	R	O	N
A	J	D	F	A	K	L	M	N	T	S	N	T

What do you do to make a baby sleep on a space ship?

You ____ ____ ____ ____ ____ ____.

Answers

LESSON 4-1

Practice A

1. 8:5
2. 13:5
3. 8:13
4. 5:8
5. 9 to 17
6. 14 to 9
7. 17 to 14
8. 9 to 14
9. $\frac{5}{8}$
10. $\frac{8}{3}$
11. $\frac{5}{3}$
12. $\frac{3}{8}$
13. 16:9
14. 7:16
15. 9:7
16. 7:9
17. $\frac{4}{12}$, 4 to 12, 4:12; or $\frac{1}{3}$, 1 to 3, 1:3
18. $\frac{12}{10}$, 12 to 10, 12:10; or $\frac{6}{5}$, 6 to 5, 6:5
19. $\frac{4}{10}$, 4 to 10, 4:10; or $\frac{2}{5}$; 2 to 5; 2:5
20. $\frac{10}{26}$, 10 to 26, 10:26 or $\frac{5}{13}$; 5 to 13; 5:13
21. 2 to 1
22. Art Club

Practice B

1. $\frac{22}{18}$, 22 to 18, 22:18; or $\frac{11}{9}$, 11 to 9, 11:9
2. $\frac{28}{22}$, 28 to 22, 28:22; or $\frac{14}{11}$, 14 to 11, 14:11
3. $\frac{18}{28}$, 18 to 28, 18:28; or $\frac{9}{14}$, 9 to 14, 9:14
4. $\frac{18}{22}$, 18 to 22, 18:22; or $\frac{9}{11}$, 9 to 11, 9:11
5. $\frac{15}{27}$, 15 to 27, 15:27, or $\frac{5}{9}$, 5 to 9, 5:9
6. $\frac{27}{18}$, 27 to 18, 27:18; or $\frac{3}{2}$; 3 to 2, 3:2
7. $\frac{18}{60}$, 18 to 60, 18:60; or $\frac{3}{10}$, 3 to 10, 3:10
8. $\frac{42}{18}$, 42 to 18, 42:18; or $\frac{7}{3}$, 7 to 3, 7:3
9. 3 to 1
10. 4 to 3
11. Glen

Practice C

1. $\frac{63}{27}$, 63 to 27, 63:27; or $\frac{7}{3}$, 7 to 3, 7:3
2. $\frac{51}{63}$, 51 to 63, 51:63; or $\frac{17}{21}$; 17 to 21, 17:21
3. $\frac{27}{51}$, 27 to 51, 27:51; or $\frac{9}{17}$, 9 to 17, 9:17
4. $\frac{27}{114}$, 27 to 114, 27:114: or $\frac{9}{38}$, 9 to 38, 9:38
5. gorillas to other apes
6. monkeys to other apes
7. gorillas to monkeys
8. monkeys to gorillas and other apes
9. other apes to all primates
10. monkeys and gorillas to other apes
11. 7:18
12. 5 to 1
13. 8th grade

Review for Mastery

1. 8
   3
   8; 3
   8; 3; $\frac{8}{3}$
2. 5
   8
   5; 8;
   5; 8; $\frac{5}{8}$
3. 3
   5
   3; 5
   3; 5; $\frac{3}{5}$
4. 8
   16
   1; 2;
   1; 2; $\frac{1}{2}$

5. 3 to 4; 3:4; $\frac{3}{4}$

6. 8 to 9; 8:9; $\frac{8}{9}$

7. 3 to 2; 3:2; $\frac{3}{2}$

8. 9 to 29; 9:29; $\frac{9}{29}$

9. $\frac{3}{5} = \frac{24}{40}$; $\frac{5}{8} = \frac{25}{40}$; 24 < 25

Mark

Challenge

1. $\frac{70}{45}$; 1.56; $\frac{115}{70}$; 1.64
2. $\frac{60}{40}$; 1.5; $\frac{100}{60}$; 1.67
3. $\frac{80}{50}$; 1.6; $\frac{130}{80}$; 1.63
4. Exercise 3; the ratio of the longer section to the shorter section is closest to equaling the ratio of the whole segment to the longer section in this exercise.

Problem Solving

1. 2 to 3; 2:3; $\frac{2}{3}$
2. 7 to 8; 7:8; $\frac{7}{8}$
3. 50 to 31; 50:31; $\frac{50}{31}$
4. 143 to 269; 143:269; $\frac{143}{269}$
5. D
6. G
7. D

Reading Strategies

1. $\frac{2}{3}$
2. 2 to 3; 2:3
3. $\frac{5}{4}$
4. 5 to 4; 5:4
5. $\frac{4}{9}$
6. 4 to 9; 4:9

Puzzles, Twisters & Teasers

1. $\frac{1}{1}$ or 1
2. $\frac{2}{3}$
3. $\frac{3}{5}$
4. $\frac{2}{5}$
5. $\frac{7}{11}$
6. $\frac{4}{7}$
7. $\frac{2}{11}$
8. $\frac{1}{11}$
9. $\frac{4}{11}$

GOOD EYESIGHT

LESSON 4-2

Practice A

1. 2; 2
2. 53; 53
3. $7 per hour
4. 14 points per game
5. 75 calories per ounce
6. $3.50 per hour
7. 45 miles per hour
8. 479 miles per hour
9. 0.15; 0.16; 16

Practice B

1. $9.50 per hour
2. 14 students per teacher
3. 192 cars per day
4. $0.12 per minute
5. 197 points per game
6. 47.6 miles per hour
7. 34.25 miles per hour
8. 512 miles per hour
9. the 16 oz jar

Practice C

1. $17.25 per hour
2. 75 calories per ounce
3. 3.8 envelopes per minute
4. 57.8 miles per hour
5. 4.8 yards per second
6. 392 miles per hour
7. 3.6 earned runs per game

8. \$0.18 per oz; \$0.17 per oz;

   $\frac{\$8.16}{48\text{ oz}}$ is the better buy.

9. \$1.29 per ft; \$1.32 per ft;

   $\frac{\$9.03}{7\text{ ft}}$ is the better buy.

10. \$3.90 per lb; \$3.95 per lb;

   $\frac{\$25.35}{6.5\text{ lb}}$ is the better buy.

11. \$22.90 per m; \$22 per m;

   $\frac{\$13.20}{0.6\text{ m}}$ is the better buy.

Review for Mastery

1. 3; 3; \$2.50
   \$2.50 per roll
2. 16; 16; 18 mi
   18 miles per gallon
3. 8; 8; \$6.50
   \$6.50 per hour
4. 2; 2; 2.5 qt; 1 lb
   2.5 quarts per pound
5. $\frac{3\text{ in.}}{1\text{ hr}}$; 3 inches per hour
6. $\frac{18\text{ students}}{1\text{ teacher}}$; 18 students per teacher

Challenge

Problem Solving

1. 60 miles per hour
2. Melanie
3. Mr. Tanney, \$8.25 per trip;
   Ms. Elmer, \$8.33 per trip
4. Eastern Middle School
5. B
6. J
7. D
8. G

Reading Strategies

1. No; it does not compare values that are in different units.
2. Yes; it compares yards to seconds.
3. It compares miles to gallons.
4. Yes
5. No; $\frac{25\text{ mi}}{1\text{ gal}}$
6. Yes
7. No; $\frac{8\text{ ft}}{1\text{ s}}$

Puzzles, Twisters & Teasers

1. K
2. O
3. I
4. M
5. A
6. N
7. C
8. O
9. A
10. R

I N A C R O A K R O O M

LESSSON 4-3

Practice A

1. $\frac{1}{4}, \frac{1}{4}$; yes
2. $\frac{2}{3}, \frac{2}{3}$; yes
3. $\frac{4}{5}, \frac{3}{4}$; no
4. $\frac{1}{2}, \frac{1}{2}$; yes
5. $\frac{5}{6}, \frac{8}{9}$; no
6. $\frac{2}{5}, \frac{2}{5}$; yes
7. $\frac{1}{3}, \frac{1}{3}$; yes
8. $\frac{2}{3}, \frac{7}{12}$; no
9. $\frac{3}{4}, \frac{3}{4}$; yes
10. $\frac{2}{3}, \frac{3}{5}$; no
11. $\frac{1}{2}, \frac{9}{20}$; no
12. $\frac{3}{5}, \frac{3}{5}$; yes

Possible answers:

13. $\frac{1}{2} = \frac{5}{10}$

14. $\frac{3}{4} = \frac{6}{8}$

15. $\frac{5}{8} = \frac{25}{40}$

16. $\frac{4}{6} = \frac{2}{3}$

17. $\frac{1}{7} = \frac{5}{35}$

18. $\frac{10}{25} = \frac{2}{5}$

Practice B

1. yes
2. yes
3. no
4. no
5. yes
6. no
7. yes
8. no
9. no
10. yes
11. yes
12. no

Possible answers:

13. $\frac{7}{9} = \frac{70}{90}$

14. $\frac{11}{12} = \frac{110}{120}$

15. $\frac{14}{15} = \frac{28}{30}$

16. $\frac{35}{55} = \frac{70}{110}$

17. $\frac{14}{10} = \frac{7}{5}$

18. $\frac{25}{18} = \frac{100}{72}$

Practice C

1. no
2. no
3. yes
4. yes
5. no
6. yes

Possible answers:

7. $\frac{7}{31} = \frac{21}{93}$

8. $\frac{24}{51} = \frac{48}{102}$

9. $\frac{6}{29} = \frac{24}{116}$

10. $\frac{14}{23} = \frac{42}{69}$

11. $\frac{17}{39} = \frac{34}{78}$

12. $\frac{25}{32} = \frac{100}{128}$

13.

CDs	2	4	10	12	28
Tapes	5	10	25	30	70

14.

Triangles	3	9	12	30	75
Circles	2	6	8	20	50

15. 20:42; 40:84

16. 30:16; 150:80

17. $\frac{10}{18}, \frac{20}{36}$

18. $\frac{48}{22}, \frac{96}{44}$

Review for Mastery

1. 7; 105; 5; 105; yes
2. 15; 90; 9; 9; 81; no
3. 60, 54, no
4. 120, 120, yes
5. 180, 180, yes

Possible answers given.

6. $\frac{6}{10} = \frac{3}{5}$

7. $\frac{10}{15} = \frac{20}{30}$

8. $\frac{18}{24} = \frac{3}{4}$

Challenge

1. a. 4 b. 6
   c. 5 d. 15
   e. 12 f. 20
2. Answers will vary. Possible answer: Find the number of triangles first. There must be at least 5 scalene triangles. Then use the ratios.
3. Answers will vary. Possible answers given.
   a. 12 b. 18
   c. 15 d. 45
   e. 36 f. 60

Problem Solving

1. No; $\frac{234}{36} = \frac{6.5}{1}$; $\frac{288}{40} = \frac{7.2}{1}$; $\frac{6.5}{1} \neq \frac{7.2}{1}$
2. Yes; $\frac{5}{8} = \frac{15}{24}$
3. $\frac{12}{15}$; 12 adults and 15 children
4. $\frac{8}{12}$; 8 cups of pineapple juice
5. C
6. J
7. D
8. G

Reading Strategies

1. $\frac{2}{3}$
2. no

3. No, because the ratios are not equal.

4. $\frac{2}{5}$

5. yes

6. Yes, because the ratios are equal.

Puzzles, Twisters & Teasers

1. U
2. H
3. I
4. T
5. S
6. Y
7. E
8. D
9. A
10. R

THEY USE THEIR HEADS

LESSON 4-4

Practice A

1. 8 and 2x
2. 9a and 42
3. 50 and 8b
4. $x = 4$
5. $z = 5$
6. $s = 6$
7. $r = 16$
8. $h = 12$
9. $d = 3$
10. $w = 3$
11. $t = 6$
12. $k = 3$
13. $f = 2$
14. $d = 21$
15. $c = 18$
16. $a = 30$
17. $k = 12$
18. $n = 15$
19. $\frac{50}{2} = \frac{150}{x}$: $x = 6$ hours

Practice B

1. $x = 14$
2. $r = 28$
3. $k = 45$
4. $z = 9$
5. $d = 66$
6. $s = 72$
7. $w = 36$
8. $t = 12$
9. $a = 24$
10. $f = 14$
11. $h = 18$
12. $c = 135$
13. $z = 60$
14. $k = 16$
15. $n = 15$
16. $m = 45$
17. $c = 6$
18. $s = 40$
19. $e = 8.8$
20. $v = 4.25$
21. $a = 11.2$
22. $7.50
23. 7 eggs

Practice C

1. $x = 21$
2. $z = 20$
3. $a = 36$
4. $r = 65$
5. $f = 82$
6. $k = 66$
7. $w = 21$
8. $t = 35$
9. $j = 85$
10. $h = 51$
11. $d = 11.9$
12. $c = 133$
13. $m = 6.3$
14. $e = 104$
15. $s = 25.9$
16. Possible answer: $\frac{40}{50} = \frac{64}{80}$
17. Possible answer: $\frac{7}{56} = \frac{9}{72}$
18. Possible answer: $\frac{15}{45} = \frac{50}{150}$
19. 675 pounds
20. 66 pounds

Review for Mastery

1. 2; 2
   2;
   54;
   3
2. 9; 9
   9
   72
   6
3. 10; 10
   10;
   60;
   15
4. $x = 12$
5. $w = 42$
6. $c = 10$
7. $\frac{8.25}{3}$; 5; $13.75
8. $\frac{90}{2}$; 225; 5 hours
9. $3.40
10. 17 minutes

Challenge

	Do quantities increase or decrease?	Proportion	Answer
1.	both increase	40 gal; 10 d	50 gallons
2.	both decrease	80 lb; 2 yd	8 pounds
3.	both increase	2.5 c; $\frac{x \text{ c}}{6}$	3.75 cups
4.	both increase	$\frac{1 \text{ in}}{60 \text{ mi}}$	5 inches
5.	both increase	$\frac{4}{\$1.50}$; x	$3.75

Problem Solving

1. 9.6, or $9.60
2. 80 grams
3. 16 inches
4. 1,015 pounds of honey
5. C
6. J
7. B
8. H

Reading Strategies

1. yes
2. no
3. yes
4. no
5. If the cross products of two ratios are equal, the ratios form a proportion.

Puzzles, Twisters & Teasers

1. A
2. L
3. L
4. T
5. H
6. E
7. F
8. A
9. N
10. S
11. L
12. E
13. F
14. T

ALL THE FANS LEFT

LESSON 4-5

Practice A

1. C
2. F
3. A
4. J
5. B
6. G
7. D
8. J
9. C
10. H
11. A
12. G
13. a. 36 inches
    b. 16; 1; 4

Practice B

1. Possible answer: Ounces; the weight of a paperback book is similar to the weight of several slices of bread, each of which is about 1 oz.
2. Possible answer: Gallons; the capacity of a large soup pot is greater than the capacity of a large milk jug, which has the capacity of 1 gal.
3. Possible answer: Feet; the length of a dining room table is similar to the length of 8 or 10 sheets of paper, each of which has a length of about 1 ft.
4. Possible answer: Tons; the weight of an elephant is similar to the weight of a few buffalos, each of which has a mass of about 1 ton.
5. 31,680 ft
6. 6.5 lb
7. 24 pt
8. 2.5 tons
9. 144 ft
10. 6 pt
11. 78 in.
12. 5 gal
13. 52 oz
14. $2\frac{1}{8}$ lb, or 2 lb 2 oz
15. 6 ft 3 in.

Practice C

1. Possible answer: Pounds; the weight of an encyclopedia is a few times greater than the weight of a bag of 3 apples, which is about 1 lb.
2. Possible answer: Miles; the distance between two cities is greater than the combined length of 18 football fields, which is about 1 mi.

3. Possible answer: Fluid ounces; the capacity of a medicine cup is similar to the combined capacity of 2 tablespoons, which is about 1 fl oz.
4. Possible answer: Inches; the length of a hamster is similar to the combined length of several small paper clips, each of which has a length of about 1 in.
5. 9200 lb
6. 1.25 mi
7. 42 pt
8. 180 in.
9. 5 gal
10. 9.25 lb
11. <
12. =
13. >
14. >
15. <
16. >
17. 98 fl oz
18. 20.5 ft, or 20 ft 6 in.

Review for Mastery

1. 4;
   divide;
   3;
   3
2. 3;
   multiply;
   3; 27;
   27
3. larger; smaller; 112
4. smaller; larger; 6
5. 96
6. 2

Challenge

1. 5280 in.
2. 50,000 thousands
3. about 36 months
4. 3,800,000 quarters
5. 84 in. longer
6. 0.08 mi
7. about 732 months
8. 136 yd

Problem Solving

1. 105 in.
2. The walrus weighs 1,700 lb more.
3. 18.875 lb, or 18 lb 14 oz
4. The 192 fl oz-container can hold 32 fl oz, or 0.25 gal more.
5. B
6. J
7. C
8. J

Reading Strategies

1. 1 yd = 3 ft
2. $\frac{1}{3} = \frac{18}{x}$
3. $x = 54$
4. $18 \text{ yd} = \frac{18 \text{ yd}}{1} \times \frac{3 \text{ ft}}{1 \text{ yd}}$
5. 18 yd = 54 ft
6. They both show that 18 yd = 54 ft.

Puzzles, Twisters & Teasers

1. T
2. D
3. N
4. H
5. W
6. C
7. E
8. O
9. K
10. Y

THEY DON'T KNOW HOW TO COOK

LESSON 4-6

Practice A

1. C
2. G
3. B
4. F
5. B
6. J
7. C
8. J
9. D
10. H
11. C
12. G
13. Benjamin; 1.8 m = 180 cm; 165 cm < 180 cm
14. Jan's; 0.9 kg = 900 g; 900 g > 830 g

Practice B

1. Possible answer: Milliliters; a cup holds much less than a large water bottle, which may hold about 1 L.
2. Possible answer: Kilograms; a small poodle has a greater mass than a

textbook, which may have a mass of about 1 kg.

3. Possible answer: Centimeters; the width of a computer screen is less than the width of a doorway, which is about 1 m.
4. Possible answer: Grams; the mass of a pencil is similar to the mass of 10 paperclips, each of which may have a mass of about 1 g.
5. 49.6 cm
6. 680 g
7. 3.8 L
8. 0.832 g
9. 76,000 m
10. 0.029 m
11. 41 L
12. 14,900 mg
13. 78 m
14. Sam's computer; 4.2 kg = 4,200 g; 4,200 g < 4,940 g
15. Elise; 1.5 m = 150 cm; 150 cm > 96 cm

Practice C

1. Possible answer: Millimeters; the thickness of a piece of spaghetti is similar to the thickness of a dime, which is about 1 mm.
2. Possible answer: liters; The capacity of a small sink might be similar to the capacity of about 5 large water bottles, each of which holds about 1 L.
3. Possible answer: Grams; the mass of an envelope is similar to the mass of a group of 10 paperclips, each of which may have a mass of about 1 g.
4. Possible answer: Kiloliters; the capacity of a gas truck is greater than the capacity of 2 large refrigerators, which may be about 1 kL.
5. 3,060 L
6. 7.5 cm
7. 0.604 kg
8. 95 cm
9. 5,600 mL
10. 0.048 g
11. =
12. >
13. <
14. >
15. <
16. >
17. Monroe. 3.5 km = 3,500 m; 2.8 km = 2,800 m; 3,500 m + 2,800 m = 6,300 m; 6,300 m < 6,400 m.
18. Tunnel A; 860 m

Review for Mastery

1. 10; 10; 150
2. 1,000; 1,000; 0.964
3. Multiply; 1,000; 4,900
4. Divide; 100; 35.31
5. 0.468
6. 690

Challenge

Masses may vary depending upon unit chosen.

Ball	Mass
Basketball	650 g
Soccer ball	453.6 g
Football	425.2 g
Volleyball	280 g
Softball	198.4 g
Baseball	148.8 g
Tennis ball	75 g
Paddleball	65.02 g
Ping pong ball	2.45 g

Problem Solving

1. Porcupine
2. 20 days
3. Midwood Middle School; 27 m
4. Less; 18 g = 0.018 kg; 0.018 kg + 0.85 kg = 0.868 kg; 0.868 kg < 1 kg
5. B
6. H
7. B
8. F

Reading Strategies

1. a smaller unit
2. multiplication
3. 100
4. 390 cm
5. a larger unit
6. division
7. 1,000
8. 0.0148 kL

Puzzles, Twisters & Teasers

1. Y
2. O
3. U
4. R
5. E
6. A
7. R
8. D
9. R
10. U
11. M
12. S

YOUR EARDRUMS

LESSON 4-7

Practice A

1. 3.15 kg per week
2. 15,840 feet per day
3. 35 pints per week
4. 8.9 feet
5. 10.6 gallons
6. 3.3 liters
7. 6.3 kilograms
8. 150 meters/minute (about 5.6 miles/hour)

Practice B

1. 0.3 grams per day
2. $8.\overline{3}$ yards per second
3. 4.5 pints per week
4. 6.6 Liters
5. 8.4 kilograms
6. 74.5 miles
7. 945 kilograms
8. $0.\overline{2}$ mi/min (about $13.\overline{3}$ mi/hr)

Practice C

1. 3.75 miles per hour
2. 1,575 gallons per hour
3. 104.13 feet per second
4. 8007 ounces
5. 11 ounces
6. 37.8 centimeters
7. 27.4 meters
8. 375 feet/min (about 4.26 mi/hr)

Review for Mastery

1. 12 inches; inches; $\frac{1 \text{ ft}}{12 \text{ in.}}$; 8
2. 16 ounces; ounces; $\frac{1 \text{ lb}}{16 \text{ oz.}}$; 3.5
3. 100 centimeters; meters; $\frac{100 \text{ cm}}{1 \text{ m}}$; 63,500
4. pounds; $\frac{1 \text{ kg}}{2.2 \text{ lb}}$; 78.18
5. 3 feet; feet; $\frac{1 \text{ yd}}{3 \text{ ft}}$; 35
6. 1000 milliliters; milliliters; $\frac{1 \text{ L}}{1000 \text{ mL}}$; 4.375

Challenge

1. 1
2. 1,000,000
3. 1,000
4. about 8.35
5. 8.35

Problem Solving

1. 1864.5 miles
2. 14.3 miles per hour
3. 3527.4 pounds
4. 7.7 pounds
5. 108.3 feet
6. 36.1 feet
7. B
8. J
9. C
10. F

Reading Strategies

1. smaller
2. larger
3. larger
4. The answer will be larger. You are changing from a larger unit of measure to a smaller one.
5. 12
6. 8
7. 96 inches

Puzzles, Twisters & Teasers

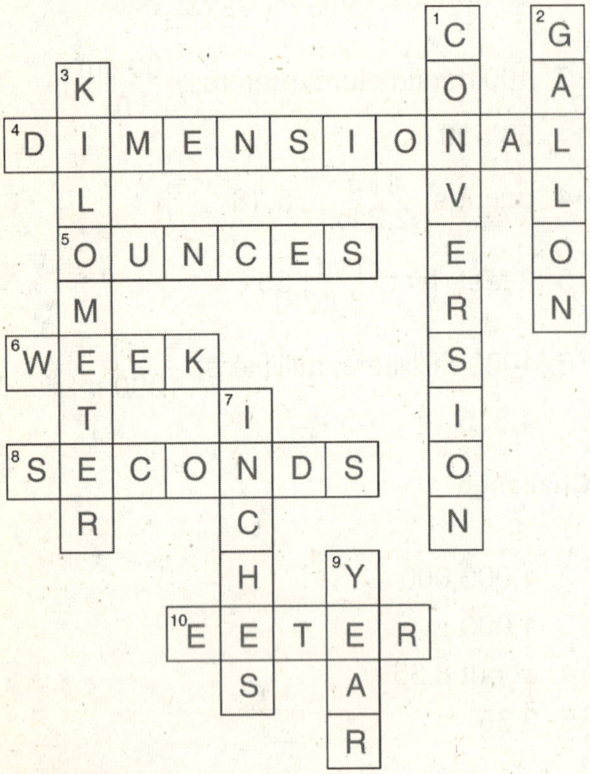

LESSON 4-8

Practice A

1. XY
2. YZ
3. XZ
4. similar; $\frac{AB}{DE} = \frac{BC}{EF} = \frac{AC}{DF}$;
   $\frac{3}{6} = \frac{4}{8} = \frac{5}{10} = \frac{1}{2}$
5. not similar; $\frac{JK}{MN} \neq \frac{KL}{NP}$;
   $\frac{5}{6} \neq \frac{7}{8}$
6. similar; $\frac{GH}{RS} = \frac{HJ}{ST} = \frac{GJ}{RT}$;
   $\frac{3}{9} = \frac{6}{18} = \frac{5}{15} = \frac{1}{3}$
7. Possible answer:
   similar; $\frac{XY}{UV} = \frac{YZ}{VW} = \frac{XZ}{UW}$;
   $\frac{4}{5} = \frac{4}{5}$
8. not similar;
   $\frac{AB}{QR} \neq \frac{BC}{RS}$; $\frac{2}{6} \neq \frac{4}{8}$
9. similar; $\frac{GH}{WX} = \frac{HJ}{XY} = \frac{JK}{YZ} = \frac{GK}{WZ}$;
   $\frac{7}{10.5} = \frac{6}{9} = \frac{4}{6} = \frac{3}{4.5} = \frac{2}{3}$

Practice B

1. not similar; $\frac{AB}{DE} \neq \frac{BC}{EF}$;
   $\frac{2}{6} \neq \frac{7}{14}$
2. similar; $\frac{RS}{UV} = \frac{ST}{VW} = \frac{RT}{UW}$;
   $\frac{3}{9} = \frac{5}{15} = \frac{1}{3}$
3. similar; $\frac{XY}{JL} = \frac{YZ}{JK} = \frac{XZ}{KL}$;
   $\frac{10}{8} = \frac{25}{20} = \frac{30}{24} = \frac{5}{4}$
4. not similar; $\frac{FG}{QR} \neq \frac{EF}{PQ} \neq \frac{EG}{PR}$;
   $\frac{13}{16} \neq \frac{15}{20} \neq \frac{16}{21}$
5. similar; $\frac{AB}{EF} = \frac{AD}{EH}$;
   $\frac{24}{36} = \frac{28}{42} = \frac{2}{3}$
6. not similar; corresponding angles are not equal

Practice C

1. similar; $\frac{XY}{RS} = \frac{YZ}{ST} = \frac{XZ}{RT}$;
   $\frac{3.2}{1.6} = \frac{6.4}{3.2} = \frac{4.8}{2.4} = 2$
2. not similar; $\frac{BC}{DE} \neq \frac{AB}{EF} \neq \frac{AC}{DF}$;
   $\frac{1.7}{3.4} \neq \frac{2.2}{5.0} \neq \frac{2.5}{5.5}$
3. not similar; $\frac{WX}{NO} = \frac{ZY}{MP} \neq \frac{WZ}{MN} \neq \frac{XY}{OP}$;
   $\frac{12}{16} = \frac{6}{8} \neq \frac{10}{15}$
4. not similar; angles are not equal
5. $BC = 6$ in.
6. $\angle M = 29°$, $\angle N = 103°$, $\angle L = 48°$
7. $FJ = 4.9$ cm

8. $\angle A = 68°$, $\angle B = 74°$,

$\angle C = 90°$, $\angle D = 128°$

Review for Mastery

1. 16; 2;

   DF; $\frac{12}{6}$; $\frac{2}{1}$;

   DE; $\frac{10}{4}$; $\frac{5}{2}$;

   no; no

2. FG; $\frac{16}{12}$; $\frac{4}{3}$;

   FH; $\frac{24}{18}$; $\frac{4}{3}$;

   GH; $\frac{20}{15}$; $\frac{4}{3}$;

   yes; yes

3. similar
4. not similar; angles not equal measure

Challenge

1.

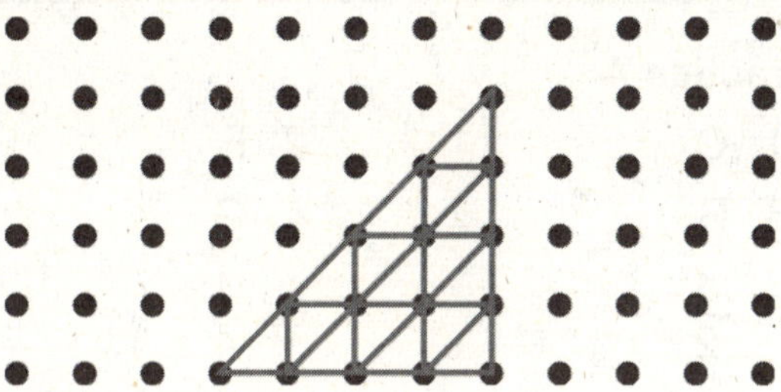

2.

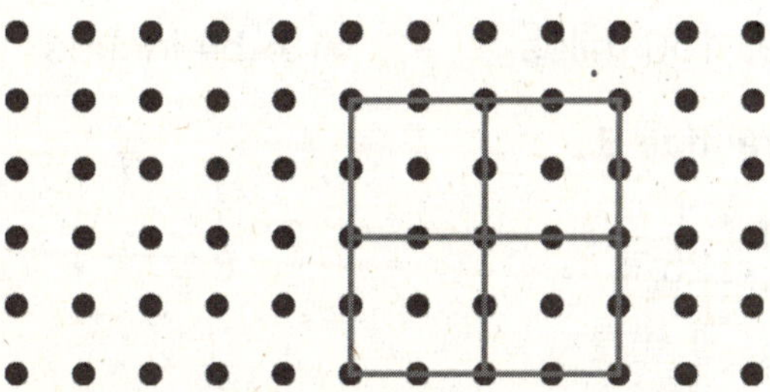

3.

4.

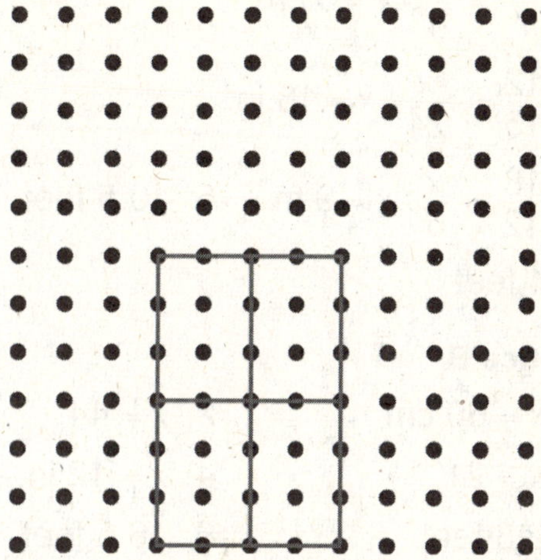

5. Drawings will vary. Possible drawing given.

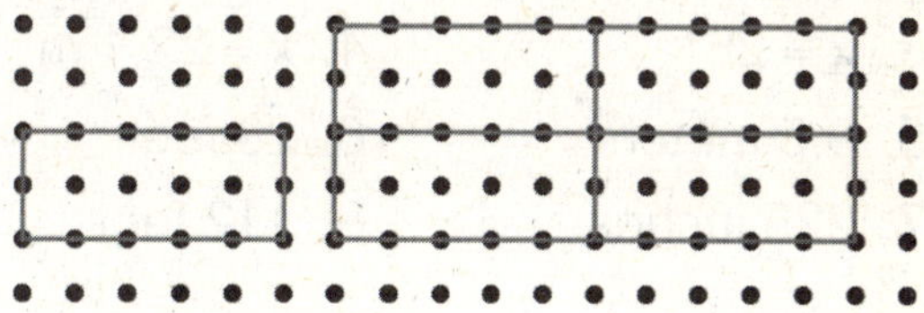

Problem Solving

1. The Dance Class
2. yes; 1:4
3. yes; 1:2
4. A
5. H
6. D
7. G

Reading Strategies

1. angle *E*
2. angle *D*
3. side *EF*
4. yes
5. no

Puzzles, Twisters & Teasers

Similar figures:

R and C; A and G; W and O;
Y and E; F and N

T U L I P S

LESSON 4-9

Practice A

1. Possible answers are given.

   $\frac{10}{5} = \frac{x}{12}$; $x = 24$ ft

2. $\frac{14}{7} = \frac{x}{4}$; $x = 8$ cm

3. $\frac{12}{16} = \frac{x}{24}$; $x = 18$ in.

4. $\frac{18}{12} = \frac{x}{6}$; $x = 9$ m   5. 10.5 feet

6. 4 feet

Practice B

1. $x = 60$ cm   2. $x = 44°$
3. $x = 21°$ 4. $x = 12$ in.
5. 2.8 feet   6. 16.5 feet

Practice C

1. $x = 21.6$ yd   2. $x = 56°$
3. $x = 26°$ 4. $x = 22.1$ m
5. 9.5 meters   6. 28 feet
7. 225 inches   8. 112 feet

Review for Mastery

1. *MO*; *MN*; *x*; 6; $x = 10$ cm   2. *AD*; *AB*; 5; $\frac{y}{15}$; $y = 27$ m
3. $k = 29°$ 4. $s = 122°$

Challenge

1. 10:15; 2:3   2. 8.5:11; 17:22
3. No, you need to leave room to print information about the dinner.
4. 2 feet by 3 feet or 1 foot by 1.5 feet
5. Yes, the ratios are equal.
6. 6 inches by 9 inches

Problem Solving

1. 78 feet long   2. 68 feet tall
3. 25 feet wide   4. 15 inches wide
5. C   6. F
7. C

Reading Strategies

1. Possible answer: because you are not actually measuring, but using proportions to find a missing length
2. Put the lengths of the sides into the proportion
3. Possible answer: $\frac{18}{6} = \frac{y}{5}$

Puzzles, Twisters & Teasers

1. PROPORTION
2. LENGTH
3. CROSS PRODUCTS
4. SOLVE
5. 25

O C T O - P U S S

LESSON 4-10

Practice A

1. C   2. F
3. B   4. G
5. $\frac{1}{2}$ 6. $\frac{1}{4}$
7. $\frac{1}{3}$ 8. $\frac{1}{7}$
9. 150 miles   10. 64 inches

Practice B

1. $\frac{1}{25}$ 2. $\frac{1}{8}$
3. $\frac{1}{9}$ 4. $\frac{1}{11}$
5. $\frac{1}{16}$ 6. $\frac{1}{9}$
7. $\frac{1}{5}$ 8. $\frac{1}{14}$
9. 35.2 feet   10. 136 miles
11. $40\frac{4}{5}$ inches

Practice C

1. $\frac{1}{4}$ 2. $\frac{1}{12}$
3. $\frac{1}{17}$ 4. $\frac{1}{21}$
5. length: 56 ft; height: 2 ft
6. length: 300 ft

7. height: 6 in.
8. height: 40 ft
9. length: 9.75 in.; height: 4.125 in.

10. 19 ft
11. 1:95

12. 150 miles
13. $193\frac{1}{5}$ inches

Review for Mastery

1. $\frac{3 \text{ in.}}{24 \text{ in.}}$; $\frac{1}{8}$
2. $\frac{4 \text{ cm}}{20 \text{ cm}}$; $\frac{1}{5}$
3. 84 inches
4. 75 miles

Challenge

	Measured Diameter	Actual Diameter	Sport
1.	2.5 cm	7.5 cm	baseball
2.	1.6 cm	24 cm	basketball
3.	3.2 cm	6.4 cm	tennis
4.	3.8 cm	3.8 cm	table tennis
5.	3.0 cm	4.2 cm	golf
6.	2.1 cm	21 cm	volleyball

Problem Solving

1. 1,100 miles
2. 1.04 feet long
3. 156.5 centimeters tall
4. 1 inch = 400 inches
5. B
6. H
7. D
8. G

Reading Strategies

1. 3 centimeters
2. Possible answer: $\frac{1}{10} = \frac{3}{x}$
3. 5 centimeters
4. Possible answer: $\frac{1}{10} = \frac{5}{x}$

Puzzles, Twisters & Teasers

V	T	S	P	R	O	P	O	R	T	I	O	N
M	N	V	S	B	R	M	K	O	L	J	I	R
O	Y	D	I	M	E	N	S	I	O	N	S	E
D	A	X	Z	R	A	T	I	O	F	F	C	P
E	O	P	E	V	O	Q	W	E	A	A	A	R
L	I	K	L	P	S	C	R	T	U	C	L	E
A	S	D	F	G	H	M	K	Q	R	T	E	S
A	C	T	U	A	L	B	V	E	Z	O	E	E
N	O	D	R	A	W	I	N	G	T	R	O	N
A	J	D	F	A	K	L	M	N	T	S	N	T

ROCKET